CPT INVARIANCE AND THE SPIN–STATISTICS CONNECTION

CPT Invariance and the Spin–Statistics Connection

Jonathan Bain

Department of Technology, Culture and Society, Tandon School of Engineering, New York University

UNIVERSITY PRESS

OXFORD
UNIVERSITY PRESS

Great Clarendon Street, Oxford, OX2 6DP,
United Kingdom

Oxford University Press is a department of the University of Oxford.
It furthers the University's objective of excellence in research, scholarship,
and education by publishing worldwide. Oxford is a registered trade mark of
Oxford University Press in the UK and in certain other countries

First Edition published in 2016

Impression: 1

Published in the United States of America by Oxford University Press
198 Madison Avenue, New York, NY 10016, United States of America

British Library Cataloguing in Publication Data
Data available

Library of Congress Control Number: 2015951929

ISBN 978–0–19–872880–1

Printed and bound by
CPI Group (UK) Ltd, Croydon, CR0 4YY

Acknowledgments

This work sprang out of several initially disparate trains of thought and coalesced in the guise of various workshop and conference presentations, and a smattering of journal articles. I would first like to thank the Department of Technology, Culture and Society at NYU-Tandon for providing me with a reduced teaching load for the semester during which the bulk of the book was written. I would also like to thank audiences at various manifestations of the International Conference on the Ontology of Spacetime, the Philosophy of Science Association Biennial Meeting, the British Society for Philosophy of Science Annual Conference, the European Philosophy of Science Association Biennial Conference, the UK and European Meeting on the Foundations of Physics, and the Annual Meeting of the Society for Exact Philosophy. The 2011 Boulder Conference on the History and Philosophy of Science allowed me to present incoherent thoughts on the CPT and spin–statistics theorems and non-relativistic field theories, as did Western University's 2009 Workshop on the Philosophy of Quantum Field Theory. I'd also like to thank the organizers and participants in the New York/New Jersey Philosophy of Science Group for providing a forum for foundational issues in the Big City. Finally, I'd like to thank Mona for all the fig bars, among other things.

Material in Chapters 1, 2, and 3 has appeared in the articles "Relativity and quantum field theory," in V. Petkov (ed.) *Space, Time, and Spacetime*, Berlin: Springer, 129–146, (2010); "CPT invariance, the spin–statistics connection, and the ontology of relativistic quantum field theories," *Erkenntnis* 78, 797–821, (2013); and "Pragmatists and purists on CPT invariance in relativistic quantum field theories," forthcoming in *European Philosophy of Science Association 2013 Conference Proceedings*. I would like to thank Springer Verlag for permission to reproduce this material. Material in Chapter 3 has appeared in the article "Quantum field theories in classical spacetimes and particles," *Studies in History and Philosophy of Modern Physics* 42, 98–106, (2011), <http://www.sciencedirect.com/science/journal/13552198>. Thanks to Elsevier for permission to include this material in the current work.

Contents

Introduction 1

Overview 5

1 The CPT and Spin–Statistics Theorems in Relativistic Quantum Field Theories 12

1.1 How to Represent CPT Invariance and the Spin–Statistics Connection 13

1.1.1 Representing CPT Invariance in RQFTs 13

1.1.2 Representing Spin and Statistics in RQFTs 15

1.2 A Plethora of Approaches 17

1.2.1 The Wightman Axiomatic Approach 18

1.2.2 Weinberg's Approach 20

1.2.3 The Lagrangian Approach 22

1.2.4 The Algebraic Approach 24

1.3 Comparison 27

1.4 Pragmatism versus Purity 29

1.4.1 Pragmatism and the Renormalization Problem 31

1.4.2 Distinguishing Pragmatism from Purity 34

1.4.3 The Existence Problem 36

1.5 Summary 38

2 The Role of Relativity in the CPT and Spin–Statistics Theorems 46

2.1 Relativity and Locality 47

2.1.1 Local Commutativity 50

2.1.2 Causality 53

2.1.3 Cluster Decomposition 57

2.2 Relativity and Geometric Modular Action 61

2.2.1 Poincaré Covariance from Modular Data 62

2.2.2 On the Fundamentality of Constraints on Modular Data 67

2.3 Relativity and CPT Invariance 71

2.3.1 An Influential But Puzzling Claim 71

2.3.2 Greenberg's Argument 72

2.4 Summary 77

3 CPT Invariance and the Spin–Statistics Connection in Non-Relativistic Quantum Field Theories 82

3.1 RQFTs versus NQFTs 84

3.2 A Plethora of Approaches to NQFTs 88

3.2.1 Axiomatic NQFTs 88
3.2.2 NQFTs in Weinberg's Approach 98
3.2.3 Lagrangian NQFTs 100
3.2.4 Algebraic NQFTs 102
3.3 Intertheoretic Relations 105
3.3.1 The Bronstein Hypercube 105
3.3.2 The Speed–Space Contraction of the Poincaré Group 109
3.3.3 A Kinematical Intertheoretic Relation 112
3.4 Summary 116

4 Non-RQFT Derivations of CPT Invariance and the Spin–Statistics Connection 120

4.1 Non-Relativistic Quantum Mechanical Derivations of the Spin–Statistics Connection 121
4.1.1 Configuration Space Approaches 121
4.1.2 Sudarshan's Approach 126
4.2 Relativistic Classical Mechanical Derivations of CPT Invariance 130
4.2.1 The Lagrangian Approach 130
4.2.2 Prospects for Alternative Approaches 133
4.3 Summary 135

5 What Explains CPT Invariance and the Spin–Statistics Connection? 139

5.1 Why CPT Invariance and the Spin–Statistics Connection Need Explanations 140
5.2 The CPT and Spin–Statistics Theorems Do Not Explain CPT Invariance and the Spin–Statistics Connection 144
5.2.1 The Deductive–Nomological Account 145
5.2.2 The Unificationist Account 147
5.2.3 Causal Accounts 148
5.2.4 The Structural Account 152
5.3 What Would Explain CPT Invariance and the Spin–Statistics Connection? 157
5.4 Summary 163

Conclusion 169

Appendix: The Separating Corollary and the Axiomatic Proof of the Spin–Statistics Theorem 173

References 177

Index 187

Introduction

This is a book on the philosophy of quantum field theory that seeks to answer the question "What explains CPT invariance and the spin–statistics connection (SSC)?" CPT invariance is the property of being invariant under the combined transformations of charge conjugation (C), space inversion (P for "parity"), and time reflection (T). It forges a link between matter and antimatter states on the one hand (which are related by charge conjugation), and spacetime symmetries on the other. The SSC is the property that holds of a state just when, if the state is characterized by Fermi–Dirac statistics, then it possesses half-integer spin, and if the state is characterized by Bose–Einstein statistics, then it possesses integer spin. It thus forges a link between the spin property of a physical system and the statistics it obeys; in particular, it provides a foundation for Pauli's exclusion principle. Physicists appeal to the CPT and spin–statistics theorems in relativistic quantum field theories (RQFTs) for an explanation of these properties, insofar as these theorems entail that any state of a physical system described by an RQFT must possess them. This appeal, however, is made problematic by the fact that there are multiple, and in some cases, mutually incompatible, ways of deriving these theorems within RQFTs. Moreover, the SSC plays an essential role in explanations of phenomena described by both non-relativistic quantum field theories (NQFTs) and non-relativistic, non-field-theoretic quantum mechanics (NQM). In these theories, the spin–statistics theorem fails, as does the CPT theorem. I will argue that these considerations entail that none of the standard accounts of scientific explanation succeeds in providing an understanding of CPT invariance and the SSC. The goal of this book is to work toward such an understanding by first providing an analysis of the necessary and sufficient conditions for these properties, and second by advocating an account of explanation appropriate for this context. Under this account, an explanation is given of a general regularity that is expressed in one type of theory (NQFTs and NQM), and that can only be adequately understood by an appeal to a more fundamental theory (RQFTs). The explanatory work is done in part by means of an appeal to intertheoretic relations, and in part by means of a derivation, within the more fundamental theory, based on a set of *non-fundamental* principles, i.e., principles that do not form an agreed-upon foundation for all formulations of the fundamental theory.

In the course of this discussion, other issues in the philosophy of quantum field theory will arise. These issues include the roles that relativity and locality play in RQFTs, the debate over which formulation of RQFTs should be adopted to address foundational issues, the role of renormalized perturbation theory in interacting RQFTs, and the nature of the intertheoretic relations between RQFTs on the one hand, and NQFTs and NQM on the other. I hope to provide clarity on these issues by viewing them through the specific lens of the CPT and spin–statistics theorems. One reoccurring feature of the discussion will be its comparison of pragmatist and purist formulations of QFTs. A received view among philosophers of physics maintains that, when it comes to addressing foundational issues associated with RQFTs, we should adopt rigorous mathematical, or as I shall call them, purist formulations of these theories. Little work has been done by philosophers in employing the methods associated with heuristic, or as I shall call them, pragmatist formulations of RQFTs. This is unfortunate since it is these latter formulations that appear in textbooks and that are used by practicing physicists to derive and test the predictions of RQFTs. While this book is not intended as a defense of pragmatism over purity, it does seek to restore some balance to the debate among philosophers.[1] As we shall see, at the end of the day, both positions face similar foundational issues.

The significance of CPT invariance and the SSC extends beyond the philosophy of quantum field theory to impact broader philosophical discussions on the nature of time reversal invariance, and the concept of indistinguishability. Note that if CPT invariance holds, then a violation of CP invariance entails a violation of T invariance.[2] Thus the CPT theorem not only entails that CPT invariance is an essential property in RQFTs, it also entails that in RQFTs in which CP invariance is violated, so too is time reversal invariance. This entailment underwrites an inference found in the physics literature from observations of CP-violation in the decay of neutral kaons to the claim that such decays are evidence for T-violation (such decay processes involve weak interactions; hence evidence that they are T-violating is evidence that the electroweak theory is T-violating). Thus explaining CPT invariance in RQFTs provides the basis for an explanation of why a significant sector of fundamental physics (i.e., electroweak theory) distinguishes the past from the future. On the other hand, it should be noted that T-violation can also be derived without the assumption of CPT invariance. Roberts (2014a) identifies two such inferences to T-violation that appear in the physics literature (one involves observations of differences in transition probabilities for neutral kaon and *B*-meson oscillations, the other involves potential observations of fundamental electric dipole states), neither of which requires the assumption of CPT invariance. Note, finally that there is a large body of recent philosophical literature on the topic of T-violation (for a summary and critique, see Roberts, 2014b), with some authors pointing out that how one interprets time reversal invariance will determine how one interprets CPT invariance (e.g., Arntzenius and Greaves, 2009; Greaves, 2010). With respect to these claims, one should note that the connection between CPT invariance and T-violation (via CP-violation) is independent of how either is interpreted.

The SSC links the statistics of a quantum state with its spin. The statistics of a quantum state underwrites a concept of indistinguishability for the physical system it characterizes. For instance, if the physical system is a collection of particles, then, in some sense, they are indistinguishable. But just how to flesh out this sense is a matter of some debate in the philosophical literature. A substantial body of work attempts to flesh out this sense in terms of the principle of the identity of indiscernibles (PII). A summary and critique of this work is given by Caulton (2013), who characterizes it in the following way: early folklore claimed that the PII holds for fermions (i.e., particles that obey Fermi–Dirac statistics), but not for bosons (i.e., particles that obey Bose–Einstein statistics), and this is because bosons can be in the same state, whereas fermions cannot. The new folklore identifies differing degrees of discernibility, and claims that one of these ("weak discernibility") allows for the PII to hold for bosons and fermions (as well as particles that obey parastatistics). The SSC adds a further dimension to this debate insofar as it connects issues concerning the identity of quantum systems, as encoded in their statistics, with the property of spin; although it remains to be seen if this connection has philosophical significance. In particular, if there were reasons to believe that the possession of spin is related to identity, independently of statistics, then explaining the SSC would provide the basis for an explanation of identity ascriptions to fermions and bosons.

Physicists tend to view the explanatory power of the SSC in more concrete terms. Statistics, we are told, characterizes how a state behaves under permutations. Again, this is easiest to think of in terms of a state that represents a collection of particles. If such a multi-particle state is invariant under a permutation of its single-particle substates, the latter can be thought of as representing indistinguishable particles. Intuitively, switching the order of any two indistinguishable particles does not affect the entire collection. Such collections of indistinguishable particles may in addition be characterized by an exclusion principle, which prohibits any two particles from being in exactly the same single-particle substate.[3] Indistinguishable particles that do not obey the exclusion principle are defined to be bosons, whereas indistinguishable particles that obey the exclusion principle are defined to be fermions. A key difference between bosons and fermions is that a collection of bosons can be prepared so that all of them are characterized by the same single-particle substate. All empirical evidence suggests that the particles described by quantum theories are either bosons or fermions.[4] Moreover, our best quantum theories of matter (as encoded in the Standard Model) represent fundamental matter states (leptons and quarks) as possessing half-integer spin and interacting *via* the exchange of integer spin carriers of fundamental gauge forces (spin-0 photons as carriers of the electromagnetic force, spin-1 W and Z bosons as carriers of the weak force, spin-1 gluons as carriers of the strong force, and the spin-1 Higgs boson that "carries" mass). The explanatory power of the SSC can now be stated: on the basis of our best theories of matter, the SSC entails that fundamental matter states must be fermions that obey the exclusion principle, whereas the states of carriers of the fundamental forces must be bosons.

These facts then provide the basis for explanations of such phenomena as lasers (described by coherent multi-photonic states) on the one hand, and Bose–Einstein condensates, superfluids and superconductors, on the other. These phenomena exhibit collective behavior that can be explained by describing them as a collection of bosons (either fundamental gauge bosons like photons or composite matter states made up of even numbers of fundamental fermions) which all start out in the same single-particle ground state. Moreover, the fact that fundamental matter states must be fermions that obey the exclusion principle provides the basis for explanations of the structure of the periodic table, and the stability of matter in general.[5]

Bose–Einstein condensates, superfluids, and superconductors are non-relativistic physical systems that are best described as possessing an infinite number of degrees of freedom and thus formulatable as NQFTs. More mundane material systems that exhibit the exclusion principle are better described in terms of finite-dimensional non-relativistic quantum mechanics (NQM). The use of NQFTs and NQM to describe such systems makes calculations more tractable, and reminds us conceptually that we are working in a domain in which relativistic effects are negligible. However, we should grant that our most fundamental descriptions of physical phenomena are framed within the context of *relativistic* QFTs. Thus one might wonder what the significance of NQFTs and NQM is to a discussion of CPT invariance and the SSC. In particular, if we know that these theories are false, given the fundamentality of RQFTs, why should it bother us that the CPT and spin–statistics theorems fail in them? In any event, there appears to be a ready explanation of this failure: these are *non-relativistic* theories! Doesn't relativity explain why CPT invariance and the SSC are essential properties in RQFTs, but only brute facts, perhaps, in NQFTs and NQM?

The short answer to this question is no! The more nuanced answer considers the following observations:

1. First, within the framework of RQFT, there are several conceptually distinct formulations of the CPT and spin–statistics theorems. These include the Wightman axiomatic approach, an approach due to Steven Weinberg, the standard textbook Lagrangian approach, and the algebraic approach.[6] These approaches not only differ on the assumptions they make, they also differ on the logical relations they impose between CPT invariance and the SSC: some approaches imply CPT invariance is necessary for the SSC, and yet others imply the SSC is necessary for CPT invariance. Moreover, while the proofs in the first three approaches assume relativity (in the sense of invariance under the restricted Lorentz group), the algebraic approach does not. In fact, as we shall see in Chapter 1, relativity is *neither* necessary *nor* sufficient for derivations of CPT invariance and the SSC in RQFTs.
2. Second, as noted above, the CPT and spin–statistics theorems fail in the context of NQFTs, but exactly why this is the case has not been made entirely clear. In the context of a particular type of NQFT, namely,

Galilei-invariant QFTs, some authors have pointed to the failure of Lorentz invariance and (relativistic) local commutativity (Lévy-Leblond, 1967, 1971), while others point to the failure of Hermiticity (Puccini and Vucetich, 2004a, 2004b). How these failures relate to each other, and whether or not they extend to NQFTs in general, remains to be made clear.

3. Finally, there is a large body of literature that attempts to derive CPT invariance and the SSC outside the framework of RQFTs. This literature is motivated in part by the examples mentioned above of non-relativistic systems for which the SSC plays an essential explanatory role. Some authors have attempted to derive the SSC within the context of NQM, while other authors have presented classical (i.e., non-quantum mechanical) derivations of the SSC, as well as CPT invariance.[7] There has been limited work by philosophers in assessing the latter attempts associated with CPT invariance (Greaves, 2008, 2010; Arntzenius and Greaves, 2009; Wallace, 2009; Arntzenius, 2011; Greaves and Thomas, 2014; Swanson, 2014), but as yet no philosophical analysis of the former body of literature on the SSC.

These considerations suggest that an explanation of the properties of CPT invariance and the SSC has yet to be made completely clear. This book aims to work toward such clarity.

Overview

The book is split into two parts: the first part (Chapters 1–4) seeks to understand the CPT and spin–statistics theorems in RQFTs, and their failure in NQFTs and NQM. The second part (Chapter 5) provides an account of explanation under which the CPT and spin–statistics theorems play a central role in explaining CPT invariance and the SSC, but not the only role.

The first part of the book, consisting of Chapters 1–4, seeks answers to the following questions:

(a) Why are CPT invariance and the SSC derivable properties in RQFTs?

(b) Why are CPT invariance and the SSC not derivable properties in NQFTs?

(c) Why are CPT invariance and the SSC not derivable properties in NQM?

The second part of the book, consisting of Chapter 5, will apply the results of the first part to answering the question,

(d) What explains CPT invariance and the SSC?

Chapters 1 and 2 address question (a). The first part of Chapter 1 reviews four distinct versions of the CPT and spin–statistics theorems in RQFTs.

These versions are based on four different approaches to formulating RQFTs: the Wightman axiomatic approach, Weinberg's approach, the Lagrangian approach, and the algebraic approach. These approaches differ not only on the principles they deem necessary to derive CPT invariance and the SSC, but also, as stated above, on how these properties are related to each other. Moreover, they also differ importantly on what they take to constitute the very notion of an RQFT. They can be grouped into two basic positions, depending on how they treat interactions: on a first gloss, pragmatist approaches trade mathematical rigor for the ability to formulate non-trivial interacting models (examples include the Weinberg and Lagrangian approaches), whereas purist approaches trade the ability to formulate non-trivial interacting models for mathematical rigor (examples include the Wightman axiomatic and algebraic approaches). Philosophers of physics are split on whether foundational questions associated with RQFTs should be framed within pragmatist or purist approaches. The last part of Chapter 1 frames this debate by viewing it through the lens of the CPT and spin–statistics theorems. The goal of this last part is to extract, from the analysis of the alternative formulations of the theorems, a more nuanced way of distinguishing between pragmatism and purity than a simple appeal to "mathematical rigor."

Chapter 2 assesses the role that relativity plays in the approaches reviewed in Chapter 1. This role takes the form of two constraints. The first constraint is restricted Lorentz invariance (RLI), i.e., invariance under the restricted (or proper orthochronous) Lorentz group $L_+^\uparrow$ (the subgroup of the Lorentz group connected to the identity that consists of Lorentz boosts but not parity or time reversal transformations, or their combination). The second constraint involves a notion of locality that, in general, prohibits influences from traveling faster than the speed of light. The approaches in Chapter 1 not only differ on how RLI is formulated (and, in fact, whether it is necessary to derive CPT invariance and the SSC), they also differ on how the locality constraint is to be understood. Three distinct notions of locality appear in these approaches: *local commutativity* in the Wightman axiomatic approach, *cluster decomposition* in Weinberg's approach, and *causality* in the Lagrangian and algebraic approaches. The first part of Chapter 2 sets about the task of unraveling the relations between these different notions of locality on the one hand, and RLI on the other. The second part of Chapter 2 considers a claim due to Greenberg (2002) which maintains that a violation of CPT invariance in an *interacting* RQFT entails a violation of RLI. This claim has been influential in both the physics and philosophy of physics literature, since it suggests a test for violations of Lorentz invariance *via* experiments that measure CPT violation (Hagar, 2009: 261; Kostelecky 2011). Moreover, in apparently linking Lorentz invariance with CPT invariance, it suggests the latter is somehow mysterious; in particular, some philosophers have wondered how the charge conjugation transformation C can arise from a purely spatiotemporal symmetry (Arntzenius and Greaves, 2009; Greaves, 2010; Arntzenius, 2011). Greenberg's claim is puzzling, however, since, as Chapter 1 demonstrates, standard proofs of the CPT theorem require more than *just* Lorentz invariance (and one doesn't even require it *at all*). Moreover, as argued in Chapter 1, such proofs hold for *non-interacting* and, at most, *unrealistic interacting* states. We will see that the distinction between pragmatist and purist approaches to the CPT theorem provides a means of addressing these puzzles and assessing Greenberg's claim.

Chapter 3 addresses question (b): "Why are CPT invariance and the SSC not derivable properties in NQFTs?" I first consider formulations of NQFTs in each of the alternative approaches to RQFTs reviewed in Chapter 1, and then assess the failure of the CPT and Spin–Statistics theorems within each approach. An essential part of this analysis involves working out the intertheoretic relation between a given formulation of RQFT and its NQFT analog. Moreover, in three of these four formulations, we will see that the violation of RLI in NQFTs plays a primary role in the failure of the CPT and spin–statistics theorems; but it is not enough to *just* say this violation explains these failures: we will see that the explicit role that the violation of RLI plays varies between approaches. Thus an adequate response to question (b) will in part have to involve identifying *how* the violation of RLI in a given approach blocks the formulation of a CPT and a spin–statistics theorem. In Section 3.1, I consider the axiomatic formulation of Galilei-invariant QFTs (GQFTs, a particular type of NQFTs) given by Lévy-Leblond (1967) and compare it with the Wightman axiomatic formulation of RQFTs. Lévy-Leblond (1967: 165) views the failure of (relativistic) *local commutativity* in GQFTs as the reason why the CPT and spin–statistics theorems fail. I suggest that this is based on an appeal to Weinberg's proofs, and is thus an illegitimate appropriation of a pragmatist result in a purist approach. I argue, instead, that the reason the CPT and spin–statistics theorems fail in axiomatic GQFTs in particular, and axiomatic NQFTs in general, is due to the differences between the complex Lorentz group on the one hand, and the complex Galilei group in particular, and in general the complex version of any isometry group associated with a classical spacetime. Such classical (viz., non-relativistic) isometry groups do not include a component connected to the identity that contains the PT transformation, as the complex Lorentz group does (the latter is an essential aspect of the axiomatic CPT and spin–statistics proofs). In Section 3.2, I consider how Weinberg's approach can be modified to describe NQFTs. In Weinberg's approach, expectation values of time-ordered products of an interaction Hamiltonian density are an essential ingredient of interacting RQFTs, and in order for such quantities to be restricted Lorentz invariant, the interaction Hamiltonian density must commute at spacelike separated distances. Since time-ordering is always invariant under the isometry group of a classical spacetime (due to the presence of an absolute temporal metric), this last requirement is unnecessary in NQFTs formulated in Weinberg's approach. I argue that this is the reason why the spin–statistics and CPT theorems fail in NQFTs in Weinberg's approach. In Section 3.3, I observe that, in the Lagrangian approach to RQFTs, the locality constraint of *causality* guarantees that time-ordered products of the relevant field quantities will be restricted Lorentz invariant. I then mount an argument similar to the one in Section 3.2 that the failure of (relativistic) *causality* in Lagrangian NQFTs is the reason why the spin–statistics and CPT theorems fail. In Section 3.4, I show that, in the algebraic approach, an essential aspect of the CPT and spin–statistics proofs is the property of vacuum separability, and that this property fails in the NQFT context. The last section 3.5 of Chapter 3 is devoted to an analysis of the intertheoretic relations between RQFTs on the one hand, and NQFTs and NQM on the other. This analysis is based on two essential distinctions. The first is a distinction between RQFTs and NQFTs in terms of spatiotemporal structure: RQFTs are QFTs in Minkowski spacetime (or, in general, Lorentzian spacetimes), while NQFTs are QFTs in classical spacetimes. The second is a distinction between theories with infinite degrees of freedom (like RQFTs and NQFTs), and theories with finite degrees of freedom (like NQM).

Chapter 4 addresses question (c): "Why are CPT invariance and the SSC not derivable properties in NQM?" It does this by evaluating attempts to derive CPT invariance and the SSC outside the framework of QFTs. Section 4.1 considers "configuration space" approaches to deriving the SSC in NQM, and an approach due to Sudarshan (1968) to deriving the SSC in NQFTs.[8] Configuration space approaches are broadly united in their attempts to derive the SSC for an NQM system with a topologically non-trivial configuration space of non-coinciding indistinguishable particles. These approaches impose indistinguishability directly on the states of a classical configuration space of particles, as opposed to encoding indistinguishability in a condition (permutation invariance) imposed on an algebra of quantum observables. Recent work in the philosophy of physics on the relation between permutation invariance and indistinguishability will shine light on the limitations of these attempts (e.g., Earman, 2010; Landsman, 2013). In particular, we shall see that "extrinsic" versions of these attempts fail to make explicit the implications of permutation invariance, appropriately understood. On the other hand, there are "intrinsic" versions of the configuration space approach that address these concerns, but while they succeed in limited cases in deriving the symmetrization postulate (the "Bose–Fermi alternative"), and, in even more limited cases, the SSC, a completely general proof of the latter has yet to be established.[9] Sudarshan's approach demonstrates that imposing $SU(2)$ invariance on a sufficiently generic non-relativistic Lagrangian that is dependent on Hermitian fields is sufficient to derive the SSC. However, as numerous authors have pointed out (Wightman, 1999, 2000; Hagen, 2004; Puccini and Vucetich, 2004a, 2004b), to the extent to which NQFTs exist that violate the SSC, any derivation of the latter for a non-relativistic field-theoretic system must go beyond the standard treatment. The discussion in Chapter 3 will indicate the sense in which Sudarshan's approach is non-standard.

Section 4.2 of Chapter 4 considers derivations of CPT invariance in relativistic classical (i.e., non-quantum mechanical) field theory. Bell (1955) demonstrated how PT invariance can be derived for relativistic classical fields. Wallace (2009) has shown how CPT invariance in the Lagrangian approach to RQFTs follows from a particular type of PT invariance in relativistic classical field theories. Greaves and Thomas (2014) generalize both of these results to obtain CPT and

PT theorems in both relativistic classical and quantum field theories in the Lagrangian approach. Their work provides answers to the following questions in the Lagrangian approach: Can CPT invariance be derived for a relativistic classical field theory? Under what conditions does PT or CPT invariance of a relativistic classical field theory entail either PT or CPT invariance of the corresponding RQFT? Under what conditions can a CPT theorem be said to be a PT theorem? After reviewing how these questions are answered in the Lagrangian approach, the remainder of Section 4.2 considers the extent to which answers to these questions can be formulated in the axiomatic, Weinberg, and algebraic approaches.

Chapter 5 addresses question (d): "What explains CPT invariance and the SSC?" I shall attempt to answer this question by breaking it into three related questions:

(i) Do the CPT and spin–statistics theorems explain CPT invariance and the SSC in RQFTs?
(ii) Why do systems described by interacting NQFTs and NQM exhibit the SSC?
(iii) Why do systems described by realistic interacting RQFTs exhibit CPT invariance and the SSC?

Questions (ii) and (iii) are motivated by the facts that the experimental evidence for CPT invariance comes from realistic interacting RQFTs, whereas that for the SSC comes, in addition, from interacting NQFTs and NQM. The received view among physicists is that the answer to question (i) is "yes" and that currently there is no feasible answer to question (ii). I shall argue, on the contrary, that the answer to question (i) is "no," at least according to current philosophical accounts of explanation, and that an explanation can be constructed that answers question (ii). This explanation explains by virtue of both a derivation from a set of non-fundamental principles in RQFTs, and an explication of intertheoretic relations between RQFTs on the one hand, and NQFTs and NQM on the other. This type of explanation is similar to one described by Weatherall (2011). On the other hand, I will conclude that, at least currently, there is no adequate explanation that can fully answer question (iii), due to the nature of a foundational problem at the heart of interacting RQFTs.

Finally, a note on presentation. This book is intended as a work in the philosophy of physics. This sub-field of philosophy tends to be fairly technical and physics-oriented, but at the same time aspires to engage with broad philosophical debates of interest to philosophers of science and generalists. The text consequently attempts to strike a balance, appropriate for philosophy of physics, between introductory exposition, technical details, and philosophical harvest. Wherever possible, technical details are left to endnotes, and chapters begin and end with summaries that seek to separate such details from their philosophical harvest.

NOTES

1. See the exchange between Fraser (2011) and Wallace (2011) for an entry into this debate.
2. See, e.g., Roberts (2014a: 3). Suppose CPT invariance is taken to be a property of a theory's Hamiltonian H in the sense that H commutes with the CPT transformation: (CPT)H = H(CPT). Then CP(THT^{-1}) = (H)CP. Thus if THT^{-1} = H (i.e., if the Hamiltonian is T-invariant), then CP(H) = (H)CP (i.e., the Hamiltonian is CP-invariant). Thus if the Hamiltonian violates CP invariance, then it violates T invariance.
3. This exclusion principle was originally formulated by Pauli in 1925 and restricted to electrons in atoms. The conjunction of the SSC and the more general exclusion principle that holds for fermions provides the basis for an explanation of Pauli's original exclusion principle for electrons.
4. In principle, a collection of indistinguishable quantum particles can be such that some subset obeys Bose–Einstein statistics while the remaining subset obeys Fermi–Dirac statistics. The collection is then said to obey "parastatistics." Baker et al. (2014) rigorize a standard argument in the physics literature that claims that any theory of "paraparticles" is a notational variant of a theory of bosons or fermions characterized by a global non-abelian gauge group.
5. These and similar explanations are considered in more detail in Chapter 5.
6. For the Wightman axiomatic approach, see Jost (1957), Burgoyne (1958), Lüders and Zumino (1958), and Streater and Wightman (1964). For Weinberg's approach, see Weinberg (1964, 1995). For the Lagrangian approach, see Fierz (1939), Pauli (1940), Kaku (1993), and Peskin and Schroeder (1995). For the algebraic approach, see Doplicher et al. (1974), Guido and Longo (1995), and Araki (1999).
7. A sample of spin–statistics attempts in NQM includes Balachandran et al. (1993), Berry and Robins (1997, 2000a, 2000b), Duck and Sudarshan (1998), Finkelstein and Rubinstein (1968), Forte (2007), Kuckert (2004), Harrison and Robbins (2004), Peshkin (2006), Reyes-Lega (2011), Reyes-Lega and Benavides (2010), Shaji (2009), Sudarshan (1968), and Tscheuschner (1991). For classical spin–statistics attempts, see Mickelsson (1984) and Morgan (2004). For classical CPT presentations, see Pauli (1940), Bell (1955), Greaves (2008, 2010), Wallace (2009), and Greaves and Thomas (2014).
8. A sample of different versions of the configuration space approach includes Finkelstein and Rubinstein (1968), Laidlaw and Dewitt (1974), Leinaas and Myrheim (1977), Tscheuschner (1991), Balachandran et al. (1993), Berry and Robbins (1997, 2000), Harrison and Robbins (2004), Peshkin (2006), Forte (2007), Reyes-Lega and Benavides (2010), and Reyes-Lega (2011).

For Sudarshan's approach, see, e.g., Sudarshan (1968), Duck and Sudarshan (1997), and Shaji (2009).

9. The intrinsic configuration space approach is developed in Bourdeau and Sorkin (1992) and elaborated in Papadopoulos et al. (2004), Benavides and Reyes-Lega (2010), Papdopoulos and Reyes-Lega (2010), Reyes-Lega and Benavides (2010), and Reyes-Lega (2011).

1

The CPT and Spin–Statistics Theorems in Relativistic Quantum Field Theories

This chapter has two goals. The first is to answer the question "Why are CPT invariance and the spin–statistics connection (SSC) derivable properties in relativistic quantum field theories (RQFTs)?" The second goal is to use the answer to this question to probe a current debate in the philosophy of quantum field theory over what version of the theory should be adopted to inform foundational issues.

Section 1.1 sets the stage by reviewing how CPT invariance and the SSC are typically represented in RQFTs. Section 1.2 then reviews four distinct versions of the CPT and spin–statistics theorems in RQFTs. These versions are based on four different approaches to formulating RQFTs: the Wightman axiomatic approach, an approach due to Steven Weinberg, the Lagrangian approach found in most textbooks, and the algebraic approach. These approaches differ not only on the principles they deem necessary to derive CPT invariance and the SSC, but also on how these properties are related to each other. Moreover, they also differ importantly on what they take to constitute the very notion of an RQFT. Section 1.3 summarizes these differences. Section 1.4 then indicates the sense in which the approaches can be grouped into two basic positions, depending on how they treat interactions: on a first gloss, *pragmatist* (or "heuristic") approaches trade mathematical rigor for the ability to derive predictions from *realistic interacting* theories ("realistic" in a sense to be made precise in Section 1.4), whereas *purist* (or "rigorous") approaches trade the ability to formulate realistic interacting theories for mathematical rigor. Philosophers of physics are split on whether foundational issues related to RQFTs should be framed within pragmatist or purist approaches. Section 1.4 frames this debate by viewing it through the lens of the CPT and spin–statistics theorems. While pragmatist and purist versions of these theorems apply unproblematically to non-interacting states, and some unrealistic interacting states, extending them to realistic interacting states is problematic: for both pragmatists and purists, to do so requires confronting foundational problems. The goal of Section 1.4 is to extract, from the analysis of the alternative formulations of the theorems in the preceding sections, a more nuanced way of

CPT Invariance and the Spin–Statistics Connection. Jonathan Bain.

distinguishing between pragmatism and purity than a simple appeal to "mathematical rigor." This will be based on the sense in which an RQFT can be said to exist.

1.1 How to Represent CPT Invariance and the Spin–Statistics Connection

Recall that CPT invariance is the property of being invariant under the combined transformations of charge conjugation C, space inversion P, and time reflection T. The SSC is the property that holds of a state just when, if the state is characterized by Fermi–Dirac statistics, then it possesses half-integer spin, and if the state is characterized by Bose–Einstein statistics, then it possesses integer spin. In order to understand how the CPT and spin–statistics theorems entail these are essential properties in RQFTs, we first need to understand how they can be represented in RQFTs.

1.1.1 Representing CPT Invariance in RQFTs

C, P, and T are transformations that act on the states of a physical system. A transformation can be understood mathematically as an element of a symmetry group, and the action of such elements on a space of states is given in terms of a representation of the group on the space. Thus in order to represent C, P, and T, we have to accomplish two tasks: first, we have to identify the symmetry groups that C, P, and T are elements of, and second, we have to identify the appropriate representations of these groups on our space of states. The second task requires us to identify the states of our physical system. While the first task is general enough to be common to all the approaches to the CPT and spin–statistics theorems reviewed in Section 1.2, how the second task is accomplished varies from approach to approach.

Note that in order to fully specify a symmetry transformation, in principle we also need to accomplish a third task; namely, we need to specify what a given representation of a symmetry transformation on our state space does to the observables associated with our physical system. As Streater and Wightman (1964: 8) indicate, simply specifying how the symmetry transformation acts on states isn't enough. Intuitively, a symmetry transformation should map states into states while preserving observables. How this third task is accomplished also varies from approach to approach.

Symmetry Groups

With respect to the first task, P, T, and PT are spacetime transformations. For RQFTs these transformations are given by elements of the Lorentz group L, the symmetry group of Minkowski spacetime (i.e., the group of transformations that leave the Minkowski metric invariant). Understood as a matrix that acts on Minkowski 4-vectors $x = (x^0, x^i), i = 1, 2, 3$, an element Λ of L belongs to one of four components, $L = L_+^\uparrow \cup L_+^\downarrow \cup L_-^\uparrow \cup L_-^\downarrow$, where $\pm$ indicates a determinant of ± 1 and $\uparrow(\downarrow)$ indicates that the sign of the 00th element $\Lambda^0{}_0$ (i.e., the element that transforms the temporal component x^0) is ± 1. Continuous (or "small") Lorentz transformations are those that can be connected to the identity and are elements of $L_+^\uparrow$, the restricted (or "proper orthochronous") Lorentz group. The discrete (or "large") transformations P, T, and PT that are not connected to the identity are elements of $L_-^\uparrow$, $L_-^\downarrow$, and $L_+^\downarrow$, respectively. Their actions on 4-vectors are defined as follows: P reverses the spatial components of a 4-vector, $\mathrm{P}(x^0, x^i) = (x^0, -x^i)$; T reverses the temporal component $\mathrm{T}(x^0, x^i) = (-x^0, x^i)$; and PT reverses both components, $\mathrm{PT}(x^0, x^i) = (-x^0, -x^i)$.

C is charge conjugation, which is supposed to map matter states to antimatter states, where the latter only differ from the former in the sign of their charge. More abstractly, according to Noether's theorem, a conserved charge is associated with an internal symmetry group (examples include the $U(1)$ symmetry associated with the electric charge of electromagnetism and the $SU(3)$ symmetry associated with the gluon charges of quantum chromodynamics). Such internal symmetries differ from theory to theory. For some theories, the representation of an internal symmetry group can be decomposed into two irreducible representations. This allows one to identify the carriers of these representations (i.e., the states they act on) as matter and antimatter states, and the map between such states as the charge conjugation transformation C.[1]

States and Observables

The second task that needs to be accomplished in order to represent C, P, and T is to identify a space of states on which the representations of our symmetry groups act. This is important insofar as identifying the states of a theory goes some way toward providing it with an interpretation. With the exception of the algebraic formalism, the approaches to the CPT and spin–statistics theorems reviewed in Section 1.2 all represent the states of an RQFT by unit rays in a Hilbert space. In the Weinberg and Lagrangian approaches, the Hilbert space is assumed to be either a Fock space of multi-particle states or a wavefunctional space of field states.[2] The Wightman approach takes field states as fundamental, but does not restrict the Hilbert space of field states to a wavefunctional space.[3] The algebraic approach abstracts even further and allows states to be complex-valued maps defined on a local algebra of observables.

For the approaches that adopt a Hilbert space of states, a theorem due to Wigner demonstrates that representations of the elements of the relevant symmetry groups take the form of unitary and antiunitary operators, call them C, P, T, on the Hilbert space. Under the assumption that the energy spectrum of the states is positive (what is known generally as the "spectrum condition"), C and P must be unitary, whereas T (and thus PT and CPT) must be antiunitary.[4] The algebraic approach is a bit more nuanced: spacetime symmetry transformations, such as P and T, are defined as maps on the local algebra of observables, whereas charge conjugation C is restricted to a map between particular representations of the algebra, viz., what are called "DHR" representations (see Section 1.2.4).

The third and final task that needs to be accomplished in order to represent C, P, and T is to identify the observables that remain invariant under the representations of these transformations on the relevant state space. At this stage, all that needs to be reported is that the various approaches reviewed in Section 1.2 differ on what they deem to be the fundamental observables associated with RQFTs. To make a rough and ready initial distinction, if the relevant "CPT observables" are identified as those objects that the CPT theorem proves to remain invariant under a CPT transformation, then the Wightman approach views them as fields, the Weinberg and Lagrangian approaches view them as encoded in a Hamiltonian density, and the algebraic approach views them as elements of a von Neumann algebra (i.e., bounded linear operators that act on a Hilbert space). More of the details of this initial sketch of ways of identifying states and observables will emerge in Section 1.2.

1.1.2 Representing Spin and Statistics in RQFTs

To represent the SSC in RQFTs, we need a way of representing the statistics associated with a state, and a way of representing the spin associated with a state.

Statistics

The statistics associated with a state characterizes how the state behaves under a permutation transformation. In all but one of the approaches reviewed in Section 1.2, this behavior is encoded in one of two ways (again, the exception is the algebraic approach):

(i) Statistics can be encoded in creation and annihilation operators a, $a^\dagger$ that act on multi-particle states in a Fock space by requiring,

$$[a(\mathbf{p}), a^\dagger(\mathbf{p}')]_\mp = \delta(\mathbf{p} - \mathbf{p}') \tag{1.1}$$

for 3-momenta $\mathbf{p}$, where "$\mp$" indicates a commutator or anticommutator, depending on whether the particle states are bosonic (i.e., obey Bose–Einstein statistics) or fermionic (i.e., obey Fermi–Dirac statistics). Creation and annihilation operators that commute will create or annihilate multi-particle states that are symmetric under a permutation of single-particle substates, whereas creation and annihilation operators that anticommute will create or annihilate multi-particle states that are antisymmetric under such a permutation. In both cases, such multi-particle states are permutation invariant.[5] In addition, the symmetric case allows, whereas the antisymmetric case does not allow, the presence of single-particle substates that agree on all their non-spatiotemporal properties (i.e., the antisymmetric case obeys Pauli's exclusion principle).

(ii) Statistics can be encoded in field operators by requiring,

$$[\phi(x), \phi^{\dagger}(y)]_{\mp} = 0, \text{ for spacelike } (x-y), \qquad \text{(StLC)}$$

where "$\mp$" indicates a commutator or anticommutator, *depending on whether the fields are bosonic or fermionic.* I will call this condition the *statistics–locality connection* (StLC). It says bosonic fields commute and fermionic fields anticommute. StLC guarantees that, when a Fock space formulation is available, the creation and annihilation operators corresponding to the fields satisfy condition (1.1). Thus, to say that a field is bosonic (resp. fermionic), could mean either that, *by definition*, the field satisfies StLC, or that, when a Fock space formulation is available, the corresponding creation/annihilation operators are associated with particle states that are bosonic (resp. fermionic).[6]

Procedure (i) suggests the bearers of statistics are particles, insofar as it encodes statistics in a way that refers explicitly to the behavior of particle states under permutations. Procedure (ii) suggests the bearers of statistics are fields, insofar as it encodes statistics in a way that refers explicitly to the behavior of fields. The algebraic approach, as we will see, eschews both procedures and instead encodes statistics in a representation of a local algebra of observables by requiring it to admit (i.e., be invariant under) a finite representation of the permutation group.

Before moving on to how spin can be represented, it should be noted that StLC is distinct from what is referred to as (relativistic) *local commutativity* (LC), which simply requires that fields either commute or anticommute at spacelike separated distances:

$$[\phi(x), \phi^{\dagger}(y)]_{\mp} = 0, \text{ for spacelike } (x-y), \qquad \text{(LC)}$$

where "$\mp$" indicates a commutator or anticommutator, *with no reference to the statistics or spin of the fields.* LC is a locality constraint that appears as an axiom in numerous axiomatic approaches to formulating RQFTs. For the sake of future discussion, a third relation should also be mentioned at this point, what I will call the *spin–locality connection* (SpLC):[7]

$$[\phi(x), \phi^{\dagger}(y)]_{\mp} = 0, \text{ for spacelike } (x-y), \qquad \text{(SpLC)}$$

where "$\mp$" indicates a commutator or anticommutator, *depending on whether the fields are integer-spin or half-integer spin.* Whereas StLC and SpLC separately entail condition (1.1) (provided a Fock space formulation is available), LC does not. LC just entails that the creation and annihilation operators associated with a given field (if they exist) either commute or anticommute. In particular, this does not determine the statistics that these operators obey, nor does it determine their spin. Evidently, both StLC and SpLC separately entail LC, but the converse is not the case. The relations between StLC, SpLC, and LC will be discussed further in Chapter 2.

Spin

The spin of a state can be understood as characterizing how the state behaves under a rotation. An integer spin state picks up a phase of $n\theta$ under a rotation by an angle θ, for some integer n, whereas a half-integer spin state picks up a phase of $(n/2)\theta$. A rotation of 2π thus changes the phase of a half-integer spin state by -1, while leaving an integer spin state unchanged. Such a 2π rotation corresponds to the identity on integer spin states, but only "half" the identity on half-integer spins states: to return to the initial half-integer spin state requires an additional 2π rotation. Thus to every action by the identity on an integer spin state, there corresponds two actions on a half-integer spin state. This behavior is entailed by the following group-theoretic facts. First, rotations in $(n+m)$ dimensions are represented by elements of the group $SO(n,m)$, generated by $(n+m)$-dimensional orthogonal matrices with unit determinant. One can show that $SO(n,m)$ has a double-cover called the "spin" group $Spin(n,m)$.[8] Carriers of representations of $Spin(n,m)$ are double-valued carriers of representations of $SO(n,m)$. This means, in particular, that under a 2π rotation, carriers of $Spin(n,m)$ change sign, whereas carriers of $SO(n,m)$ do not. In general, then, an integer spin state in $(n+m)$ dimensions can be identified as a carrier of a representation of $SO(n,m)$, whereas a half-integer spin state in $(n+m)$ dimensions can be identified as a carrier of a representation of $Spin(n,m)$.

In the case of RQFTs, we're concerned with spin states in (3+1) dimensions. The relevant "rotation" group is then the restricted Lorentz group $L_+^\uparrow$, and its corresponding double covering spin group is $SL(2,\mathbb{C})$, the group generated by 2×2 complex matrices with unit determinant. Thus, in RQFTs, integer spin states can be identified as carriers of representations of $L_+^\uparrow$, while half-integer spin states can be identified as carriers of representations of $SL(2,\mathbb{C})$. As alluded to above, identifying the states of a physical system as carriers of a representation of a symmetry group is not enough to fully specify what such states are, i.e., in terms of the discussion above, whether such states are multi-particle states, field states, or states associated with a local algebra of observables.

1.2 A Plethora of Approaches

Now that we have reviewed *what* CPT invariance and the SSC *are*, and how they can be represented in RQFTs, we can move on to the sense in which the CPT and spin–statistics theorems assert that they are *essential* properties in RQFTs. This will involve reviewing four distinct approaches to formulating these theorems: the

Wightman approach, Weinberg's approach, the standard Lagrangian approach, and the algebraic approach. As alluded to in Section 1.1, each of these approaches can be associated with a distinct way of understanding what an RQFT is about, i.e., what the basic objects of an RQFT are, and what principles these basic objects are supposed to satisfy. The immediate aim of this review is to explicitly identify the assumptions that underlie each approach in order to facilitate comparison in the following sections.

1.2.1 The Wightman Axiomatic Approach

In the Wightman axiomatic approach, the CPT theorem was derived originally by Jost (1957) and the spin–statistics theorem by Lüders and Zumino (1958) and Burgoyne (1958). The basic objects are vacuum expectation values of unordered products of fields, referred to as Wightman functions, $W^{(n)}(x_1, \ldots, x_n) \equiv \langle 0 \,|\, \phi(x_1) \ldots \phi(x_n) \,|\, 0 \rangle$, where $\phi(x)$ is a generic quantum field (technically defined as an operator-valued distribution), and $|0\rangle$ is its vacuum state. Wightman functions are required to satisfy a number of axioms, and it is the goal of this approach to construct models of these axioms that represent interacting RQFTs.[9] For the purposes of deriving CPT invariance and the SSC, one first adopts the following two assumptions:

(i) *Restricted Lorentz invariance* (RLI). The fields that appear in Wightman functions are invariant under the restricted Lorentz group $L_+^\uparrow$.

(ii) *Spectrum condition* (SC). The fields possess positive energy, in the sense that the spectrum of the momentum operator associated with $L_+^\uparrow$ is confined to the forward light cone.

Assumptions (i) and (ii) entail that Wightman functions can be analytically extended to complex-analytic functions that are invariant under the proper complex Lorentz group $L_+(\mathbb{C})$ which, unlike $L_+^\uparrow$, contains parity and time reversal transformations. Invariance under $L_+(\mathbb{C})$ entails complex Wightman functions obey the following PT invariance property:

$$W^{(n)}(\zeta_1, \ldots, \zeta_{n-1}) = (-1)^{\mathcal{J}} W^{(n)}(-\zeta_1, \ldots, -\zeta_{n-1}) \qquad \text{(PT)}$$

where $\zeta_i = \xi_i - i\eta_i$, with $\xi_i, \eta_i \in \mathbb{R}$, and $\xi_i = x_{i+1} - x_i$ are difference variables. $\mathcal{J}$ is the number of (conjugate) spinor components in all fields occurring in $W^{(n)}$. Intuitively, $\mathcal{J}$ encodes the spin of these fields: if $\mathcal{J}$ is even, the fields have integer spin and if $\mathcal{J}$ is odd, they have half-integer spin. The invariance property PT is a special case of invariance under $L_+(\mathbb{C})$; namely, it is invariance under parity and time-reversal. As Wightman (1999: 744) observes, this is a remarkable fact: "... although only invariance under $L_+^\uparrow$ was assumed for the n-point vacuum expectation values, the associated analytic function is invariant under spacetime inversion."

Now suppose we impose the wrong spin–statistics connection on the fields. To do this, we first encode statistics in the fields by assuming the StLC of Section 1.1.2, i.e., we assume bosonic fields commute and fermionic fields anticommute. We next assume the *wrong* spin–statistics connection, i.e., we assume half-integer spin fields are bosonic and integer spin fields are fermionic. Call this NSSC, where SSC is the spin–statistics connection. The conjunction of StLC and NSSC entails half-integer-spin fields commute and integer-spin fields anticommute, i.e., it entails the *wrong spin–locality connection*, call this NSpLC, where SpLC is the spin–locality connection of Section 1.1.2. We thus have in the first instance

(a) [StLC & NSSC] ⇒ NSpLC

One can now show that the conjunction of PT and NSpLC entails that the fields annihilate the vacuum (see Appendix). Moreover, one can show that if the fields annihilate the vacuum, and if they satisfy LC, then they must vanish identically.[10] Since RLI and SC entail PT, and since StLC entails LC, we have the following implication,

(b) [RLI & SC & StLC & NSSC] ⇒ (fields vanish)

or: non-vanishing relativistic positive-energy fields that encode statistics via StLC must be characterized by ~NSSC. If we assume this is equivalent to SSC, i.e., if "... one puts aside the possibility of laws of statistics other than Bose–Einstein or Fermi–Dirac" (Streater and Wightman, 1964: 147), then non-vanishing fields must possess the SSC.[11]

Now suppose that instead of StLC, the fields obey *weak local commutativity*:

(iii) *Weak local commutativity* (WLC). At (or in the neighborhood of) a Jost point[12] the fields satisfy $\langle \mathbf{0} | \phi(x_1) \ldots \phi(x_n) | \mathbf{0} \rangle = i^F \langle \mathbf{0} | \phi(x_n) \ldots \phi(x_1) | \mathbf{0} \rangle$, where F is the number of fermionic fields.[13]

Jost (1957) showed that the conjunction of (i), (ii), and (iii) entails the existence of an antiunitary operator that combines the actions of C, P, and T transformations on fields, leaving them invariant (Streater and Wightman, 1964: 146).[14]

To recap, in the axiomatic approach, we have the following schematic entailments:[15]

A1. [(RLI for fields) & SC & StLC] ⇒ (SSC for non-vanishing fields)

A2. [(RLI for fields) & SC & WLC] ⇒ (CPT invariance of fields)

1.2.2 Weinberg's Approach

Weinberg's (1964, 1995) approach considers an RQFT to be a theory about particles that possess conserved charges, and that interact *via* scattering events described by an S-matrix.[16] One begins with the definition of single-particle states as finite irreducible representations of the restricted Lorentz group and builds quantum fields out of them in order to construct an S-matrix with two essential properties. Weinberg (1964: 1318) originally based this approach on three assumptions:

(i) *Perturbation theory*. The S-matrix is given by a power series expansion in time-ordered products of an interaction Hamiltonian density $\mathfrak{H}_{int}(x)$:

$$S_{\beta\alpha} = \sum_{n=0}^{\infty} \frac{-i^n}{n!} \int \langle \beta \,|T\{\mathfrak{H}_{int}(x_1)\ldots\mathfrak{H}_{int}(x_n)\}|\,\alpha\rangle d^4x_1 \ldots d^4x_n \qquad (1.2)$$

where $|\beta\rangle, |\alpha\rangle$ are asymptotic multi-particle states, and the time-ordered product $T\{\mathfrak{H}_{int}(x_1)\ldots\mathfrak{H}_{int}(x_n)\}$ orders the $\mathfrak{H}_{int}(x_i)$ according to $t_1 > \ldots > t_n$.[17]

(ii) *RLI of the S-matrix*. The S-matrix is invariant under restricted Lorentz transformations. A sufficient condition for this is that $\mathfrak{H}_{int}(x)$ is a Lorentz scalar, and it commutes at spacelike distances: $[\mathfrak{H}_{int}(x), \mathfrak{H}_{int}(y)] = 0$, for spacelike $(x-y)$.[18]

(iii) *Particle interpretation*. $\mathfrak{H}_{int}(x)$ is constructed out of the creation and annihilation operators a, $a^{\dagger}$ for free particles.

In Weinberg (1995: 177), assumption (iii) is replaced by an additional constraint on the S-matrix:

(iii′) *Cluster decomposition of the S-matrix* (CD). Let $S_{\beta_1+\ldots+\beta_N,\alpha_1+\ldots+\alpha_N}$ represent the S-matrix for N multi-particle processes $|\alpha_1\rangle \to |\beta_1\rangle, \ldots, |\alpha_N\rangle \to |\beta_N\rangle$. If all particles in states $|\alpha_i\rangle, |\beta_i\rangle$ are at a great spatial distance from all particles in states $|\alpha_j\rangle, |\beta_j\rangle$, for $i \neq j$, then the S-matrix factorizes: $S_{\beta_1+\ldots+\beta_N,\alpha_1+\ldots+\alpha_N} = S_{\beta_1\alpha_1}\ldots S_{\beta_N\alpha_N}$.

CD is a locality constraint that requires scattering experiments in regions of spacetime that are separated by a large spatial distance to be independent of each other. A sufficient condition for this is (iii).[19] The creation and annihilation operators a, $a^{\dagger}$ in (iii) are defined by their actions on multi-particle states in a Fock space constructed out of single-particle states. This definition entails that they satisfy condition (1.1) of Section 1.1.2. Weinberg then demonstrates that a sufficient condition ("the only known way" Weinberg, 1964: 1318) for (iii) to be compatible with (ii) is that $\mathfrak{H}_{int}(x)$ be a functional of free fields $\phi(x)$ characterized by the following properties:[20]

(α) $\phi(x) = \kappa\phi^{+}(x) + \lambda\phi^{-}(x)$, where $\phi^{+}(x)$ and $\phi^{-}(x)$ are linear combinations of a, $a^{\dagger}$, respectively.

(β) $\phi(x)$ satisfies RLI.

(γ) $\phi(x)$ satisfies LC.

Property (α) entails that the S-matrix satisfies CD, while properties (β) and (γ) entail that the S-matrix is RLI.[21] Property (β) is secured by an appropriate choice of the expansion coefficients in the expressions for $\phi^{+}(x)$ and $\phi^{-}(x)$, while property (γ) is secured by an appropriate choice of the constants κ, λ. Weinberg (1995: 238) then demonstrates that the choice of κ and λ that guarantees LC is necessary and sufficient for the fields to possess the SSC. Thus in the first instance we have

(a) [(RLI of S-matrix) & (CD of S-matrix)] $\Leftrightarrow$ [Properties (α), (β), (γ)]

(where the "$\Rightarrow$" is assumed on the basis of Weinberg's intuition!) and in the second instance,

(b) [Condition (1.1) & LC] $\Leftrightarrow$ SSC

Thus, for Weinberg, the SSC is a property derived for fields satisfying properties (α)–(γ). However, the existence of such fields is a consequence of the fundamental assumptions (i)–(iii) above. In particular, fields in Weinberg's approach are instrumental devices introduced to secure the properties (ii) and (iii′) of the S-matrix, and the S-matrix ultimately characterizes the behavior of multi-particle states. Thus, arguably, in Weinberg's approach, the fact that fields possess the SSC is a reflection of the more fundamental fact that particle states possess the SSC.[22]

Weinberg (1995: 198–199) further demonstrates that if the fields carry a non-zero value of a conserved charge, then to every particle state there must correspond an antiparticle state.[23] Note that this is not quite a demonstration of CPT invariance. To prove the latter, from Weinberg's point of view, requires an explicit demonstration that the full Hamiltonian density is invariant with respect to the composition of C, P, and T operators (Weinberg, 1995: 244–246). The demonstration ultimately rests on the transformation properties of the creation and annihilation operators a, $a^{\dagger}$ under C, P, and T separately, which determine how the fields transform, and hence how the Hamiltonian density transforms. The P and T transformations follow from the behavior of a, $a^{\dagger}$ under restricted Lorentz transformations, while the C transformation is posited to hold between a, $a^{\dagger}$ on the one hand, and antiparticle operators a^c, $a^{c\dagger}$ on the other, where the existence of the latter is entailed by the existence of antiparticles.

To recap, in Weinberg's approach, we have the following schematic entailments:

B1. [(RLI for S-matrix) & CD & (1.1)] ⇒ [SSC for particles]

B2. [(RLI for S-matrix) & CD & (existence of conserved charges)] ⇒ [CPT invariance of $\mathfrak{H}(x)$]

1.2.3 The Lagrangian Approach

In Weinberg's approach, one begins with the definition of single-particle states as irreducible representations of the restricted Lorentz group, and then constructs quantum fields out of them (i.e., one constructs fields as linear combinations of Fock space creation and annihilation operators). In textbook accounts of RQFTs, one begins with a classical Lagrangian field theory, and then quantizes the fields to obtain quantum fields. Field quantization involves the construction of a Fock space from the properties of the solution space of a classical field equation. Part of this process involves positing LC for the field. As Section 1.1.2 explained, this entails that the Fock space creation and annihilation operators associated with the field (if they exist) *either* commute *or* anticommute. For a field of a given spin, one then must choose between these two options, and this choice then determines the statistics of the field via condition (1.1).[24] Note that this is a reversal of Weinberg's procedure, in which LC (or StLC) is derivative, in part, of condition (1.1). To derive the SSC (i.e., to determine the choice of statistics for a field of a given spin), the textbook approach, following Fierz (1939) and Pauli (1940), introduces a causality constraint (sometimes referred to as "microcausality"):

The observable quantities associated with an RQFT commute at spacelike distances. (Causality)

Pauli (1940: 721) justifies this constraint in the following manner:

> We shall, however, expressively postulate in the following *that all physical quantities at finite distances exterior to the light cone (for* $|x_0' - x_0''| < |\mathbf{x}' - \mathbf{x}''|$*) are commutable...* The justification for our postulate lies in the fact that measurements at two space points with a spacelike distance can never disturb each other, since no signals can be transmitted with velocities greater than that of light.

Note that *causality* is a locality constraint similar to LC; in particular, both do not refer to statistics or spin. On the other hand, *causality* is distinct from LC insofar as it only requires commutativity. How *causality* relates to the SSC will depend on how the notion of an observable is to be understood. This does not appear explicitly in the textbook literature, but one can infer that the standard approach implicitly assumes a connection between observables and *spin*.

In particular, it is assumed that observables are integer-spin states or bilinears in half-integer-spin states, whereas unobservables are half-integer-spin states, as Sterman (1993) explains in the context of spin-1/2 Dirac fields:[25]

> If the commutators of Dirac fields do not vanish at spacelike distances, what becomes of causality? We recall, however, that spinors are double-valued representations of the rotation group. As such, a spinor is not itself directly observable, since a rotation by 2π changes its sign. On the other hand, operators that are bilinear in the field—such as components of the energy-momentum tensor—do not change sign, and are observables. More generally, we may consider any operator of the form $B_i(x) = \bar{\psi}(x)O_i\psi(x)$, where O_i is some matrix, possibly combined with differential operators. We can easily show that equal-time commutators between the B_i vanish, if the fields obey [anticommutation relations]. (Sterman, 1993: 167)

The textbook approach then justifies the SSC, i.e., the choice of Bose–Einstein statistics for integer-spin fields, and Fermi–Dirac statistics for half-integer-spin fields, in the following way. We first encode statistics in the particles by assuming condition (1.1) of Section 1.1.2, i.e., we assume bosonic creation and annihilation operators, a, $a^\dagger$, commute and fermionic a, $a^\dagger$ anticommute. We next assume the NSSC, i.e., we assume half-integer spin a, $a^\dagger$ are bosonic and integer spin a, $a^\dagger$ are fermionic. This entails half-integer spin a, $a^\dagger$ commute and integer-spin a, $a^\dagger$ anticommute. As in Section 1.2.1, call this NSpLC, the *wrong spin–locality connection*, here applied to creation and annihilation operators. We thus have in the first instance[26]

(a) [Condition (1.1) & NSSC] $\Rightarrow$ NSpLC

One can now show that imposing NSpLC on the creation and annihilation operators of a restricted Lorentz invariant integer spin field violates *causality*, and imposing NSpLC on the creation and annihilation operators of a restricted Lorentz invariant half-integer-spin field violates either *causality* or the SC (see, e.g., Kaku, 1993: 87, 90; Peskin and Schroeder, 1995: 52–58).[27] Thus:

(b) [RLI & (NSpLC for integer-spin a, $a^\dagger$)] $\Rightarrow$ ~*Causality*
(c) [RLI & (NSpLC for half-integer-spin a, $a^\dagger$)] $\Rightarrow$ (~*Causality* $\vee$ ~SC)

Thus relativistic integer-spin fields that encode statistics on particles via condition (1.1) must possess ~NSSC on pain of violating *causality*, and relativistic half-integer-spin fields that encode statistics on particles via condition (1.1) must possess ~NSSC on pain of violating either *causality* or the SC. As in Section 1.2.1, if ~NSSC is equivalent to the SSC, then relativistic integer-spin fields that encode statistics via condition (1.1) and satisfy *causality* must possess SSC, and relativistic half-integer-spin fields that encode statistics via condition (1.1) and satisfy *causality* and the SC must possess SSC.

In the textbook approach, CPT invariance requires a demonstration that the Hamiltonian density $\mathfrak{H}(x)$ associated with the Lagrangian density of an RQFT is invariant under the operation CPT. As outlined by Kaku (1993: 120–123), this requires two assumptions:[28]

(i) The Lagrangian density $\mathcal{L}(x)$ is a local, normal-ordered Hermitian Lorentz scalar.

(ii) The SSC holds for the fields that appear in $\mathcal{L}(x)$.

In assumption (i), a *local* Lagrangian density is a functional of field variables and their (first and possibly higher order) derivatives evaluated at the same spacetime point, i.e., $\mathcal{L}(x) = \mathcal{L}[\phi_i(x), \partial_\mu\phi_i(x)]$, for a given number i of fields. A *normal-ordered* Lagrangian density is one in which the field variables have been decomposed into positive and negative frequency parts (corresponding to creation/annihilation operators on a Fock space), and all positive frequency parts occur to the right of all negative frequency parts.[29] Finally, a *Hermitian* Lagrangian density is identical to its Hermitian conjugate: $\mathcal{L}(x) = \mathcal{L}^\dagger(x)$.

The proof involves first determining the CPT transformation properties for the types of fields that can appear in $\mathcal{L}(x)$. A primary result of this analysis is that an even-rank tensor field transforms under CPT into its Hermitian conjugate, whereas an odd-rank tensor field transforms into the negative of its Hermitian conjugate (for tensor fields built out of normal-ordered products, this result requires the SSC to hold). This entails that a Lorentz scalar constructed by contracting a set of even- and/or odd-ranked tensor fields (and/or their derivatives) transforms into its Hermitian conjugate under CPT: any minus signs due to odd-ranked tensor fields are canceled, since there must be an even number of such fields in order to construct a scalar. If the Lorentz scalar is Hermitian, it will thus be CPT invariant.

The requirement that $\mathcal{L}(x)$ be local rules out Lagrangian densities that contain an infinite number of derivative terms, which would place doubt on the claim that $\mathcal{L}(x)$ is a contraction of an even number of tensor indices.[30]

To recap, in the standard textbook Lagrangian approach, we have the following schematic entailments:

C1. [(RLI of fields) & SC & (1.1) & *causality*] $\Rightarrow$ [SSC for fermionic fields]

C2. [(RLI of fields) & (1.1) & *causality*] $\Rightarrow$ [SSC for bosonic fields]

C3. [(SSC) & (RLI of fields) & (local Hermitian normal-ordered $\mathcal{L}(x)$)] $\Rightarrow$ [CPT invariance of $\mathfrak{H}(x)$]

1.2.4 The Algebraic Approach

A formulation of the CPT and spin–statistics theorems was given by Guido and Longo (1995) in the context of algebraic quantum field theory.[31] The basic object in this approach is a net of von Neumann algebras, $\mathcal{O} \mapsto \mathfrak{R}(\mathcal{O})$, that assigns a local algebra of observables $\mathfrak{R}(\mathcal{O})$ to every double-cone region $\mathcal{O}$ of Minkowski spacetime (a double-cone region is the intersection of the causal future of a point with the causal past of another point to the future of the first). The local algebras are required to satisfy isotony: if $\mathcal{O}_1 \subset \mathcal{O}_2$, then $\mathfrak{R}(\mathcal{O}_1) \subset \mathfrak{R}(\mathcal{O}_2)$; and this entails that they generate a quasi-local algebra $\mathfrak{R}$. The elements of $\mathfrak{R}$ can be represented as bounded linear operators that act on a separable Hilbert space $\mathcal{H}_0$ with a cyclic vacuum vector Ω.[32]

Within this framework, Guido and Longo demonstrated that CPT invariance and the SSC follow from three additional assumptions. The first is the algebraic analog of *causality*, call it *algebraic causality*, which requires observables associated with spacelike separated regions to commute:

For $A_1 \in \mathfrak{R}(\mathcal{O}_1), A_2 \in \mathfrak{R}(\mathcal{O}_2)$, *and* $\mathcal{O}_1, \mathcal{O}_2$ *spacelike separated*, $[A_1, A_2] = 0$. (Algebraic causality)

Among other things, *algebraic causality* entails that the vacuum is separating (i.e., if $A\Omega = 0, A \in \mathfrak{R}$, then $A = 0$). This separating property is important in underwriting the second assumption, namely *modular covariance* (MC). This requires that, for any wedge region W of Minkowski spacetime, the *modular operators* Δ_W^{it} of the local algebra $\mathfrak{R}(W)$ of the wedge implement Lorentz boosts on $\mathfrak{R}$.[33] Formally, for any wedge W and any double cone $\mathcal{O}$,

$$\Delta_W^{it} \mathfrak{R}(\mathcal{O}) \Delta_W^{-it} = \mathfrak{R}(\Lambda_W(t)\mathcal{O}) \qquad \text{(MC)}$$

where $\Lambda_W(t)$ is the one-parameter group of Lorentz boosts that leave W invariant. MC is motivated by a theorem due to Bisognano and Wichmann (1976) which demonstrates that for a von Neumann algebra generated by local polynomial algebras of Wightman fields, the modular operators of the local algebra of any wedge implement Lorentz boosts, and the modular conjugate operator is given by the CPT operator that leaves the Wightman fields invariant.

Guido and Longo's third assumption is *additivity*, which requires that any local algebra $\mathfrak{R}(\mathcal{O})$ associated with a double cone region $\mathcal{O}$ be generated by the local algebras associated with any of its open coverings:

$$\mathfrak{R}(\mathcal{O}) = \bigcup_i \mathfrak{R}(\mathcal{O}_i), \text{ for } \mathcal{O} = \bigcup_i \mathcal{O}_i. \qquad \text{(Additivity)}$$

Additivity is assumed by Guido and Longo (1995: 529) in order to make use of Doplicher and Roberts (1990) reconstruction of the field algebra net $\mathcal{O} \mapsto \mathfrak{F}(\mathcal{O})$ associated with the observables net $\mathcal{O} \mapsto \mathfrak{R}(\mathcal{O})$.[34] The field algebra is an extension of $\mathfrak{R}(\mathcal{O})$ to include internal gauge symmetries, and thus charges and statistics (the latter, insofar as permutation symmetry is a type of gauge symmetry).

Under the above three assumptions, Guido and Longo were able to show that CPT invariance holds for a particular type of representation of $\mathfrak{R}$, what are called *DHR representations*; and that the SSC holds for a subset of such representations.

DHR representations were defined by Doplicher, Haag, and Roberts (1971: 200) in the following way:

DHR Representation

Let $(\mathcal{H}_0, \pi_0)$ be the vacuum representation of $\mathfrak{R}$ generated by the vacuum state ω_0.[35] Then a DHR representation with respect to ω_0 is a representation $(\mathcal{H}, \pi)$ such that $\pi|_{\mathfrak{R}(\mathcal{O}')}$ is unitarily equivalent to $\pi_0|_{\mathfrak{R}(\mathcal{O}')}$ for any double cone $\mathcal{O}$.

A DHR representation is unitarily equivalent to the vacuum representation except for some bounded region of spacetime $\mathcal{O}$ (where $\mathcal{O}'$ is the causal complement of $\mathcal{O}$). Thus the states associated with DHR representations are supposed to represent localized states in so far as they differ from the vacuum only in some bounded region of spacetime. Doplicher et al. (1971) showed that DHR representations possess conjugates (which can be interpreted as representing antimatter), and admit representations of the permutation group S_N (i.e., the group of permutations of N objects), thus they can be characterized in terms of statistics. DHR representations that admit finite representations of S_N are referred to as possessing finite statistics. In particular, a DHR representation that admits the trivial representation of S_N is called bosonic, and a DHR representation that admits the alternating representation of S_N is called fermionic.[36] This amounts to the algebraic way of encoding statistics. Formally, one can define the statistical phase κ_ρ of a DHR representation ρ in terms of the dimension of the representation of S_N it admits, with $\kappa_\rho = +1$ (resp. -1) characterizing the trivial (resp. alternating) representation. One can then define a "statistics" operator $\Gamma = \oplus_\rho \kappa_\rho I_{\mathcal{H}_\rho}$, where the Hilbert space $\mathcal{H}_0$ associated with $\mathfrak{R}$ decomposes into the direct sum $\mathcal{H}_0 = \oplus_\rho \mathcal{H}_\rho$ of Hilbert spaces $\mathcal{H}_\rho$ associated with DHR representations ρ. Statistics can now be encoded in the net of field algebras $\mathcal{O} \mapsto \mathfrak{F}(\mathcal{O})$ by imposing *normal commutation relations* (Guido and Longo, 1995: 524):

$$\text{Let } \mathfrak{F}_\pm(\mathcal{O}) := \{A \in \mathfrak{F}(\mathcal{O}) \,|\, \Gamma A \Gamma = \pm A\}. \text{ Then, if } \mathcal{O}_1 \text{ and } \mathcal{O}_2 \text{ are spacelike separated, then } [\mathfrak{F}_+(\mathcal{O}_1), \mathfrak{F}_+(\mathcal{O}_2)]_- = 0 = [\mathfrak{F}_-(\mathcal{O}_1), \mathfrak{F}_-(\mathcal{O}_2)]_+. \quad \text{(NCR)}$$

Normal commutation relations (NCR) are the algebraic way of enforcing the StLC of Section 1.1.2 on the elements of a field algebra. One can now demonstrate that *algebraic causality* and MC entail the existence of a positive energy unitary representation U_ρ of the covering $\tilde{P}_+^\uparrow$ of the restricted Poincaré group that acts on a given DHR representation ρ. Moreover, this representation can be extended to a representation of the covering $\tilde{P}_+$ of the proper Poincaré group, which, recall, contains spacetime reflections. The CPT theorem then amounts to a demonstration that there exists an operator Θ on DHR representations that implements spacetime reflections and intertwines a given representation with its conjugate (Guido and Longo, 1995: 530).[37] Now let $U_\rho(2\pi)$ denote an implementation of a rotation of 2π. Then for irreducible ρ, $U_\rho(2\pi) = \pm 1$, depending on whether U_ρ is characterized by integer-spin or half-integer-spin, respectively (recall that half-integer-spin representations of $\tilde{P}_{+}^{\uparrow}$ change their phase by –1 under a 2π rotation, whereas integer-spin representations do not). The spin–statistics theorem now amounts to a demonstration that the statistical phase of a DHR representation coincides with its "spin parameter," i.e., $\kappa_\rho = U_\rho(2\pi)$ (Guido and Longo, 1995: 531).

To recap, in Guido and Longo's algebraic approach, we have the following entailments:[38] For a von Neumann algebra $\mathfrak{R}$ of local observables with a cyclic vacuum representation,

D1. [*Algebraic causality* & *additivity* & MC] ⇒ [CPT invariance for DHR representations]

D2. [*Algebraic causality* & *additivity* & MC & NCR] ⇒ [SSC for irreducible, Poincaré-covariant DHR representations with finite statistics]

In D2, Poincaré covariance refers to invariance under the universal covering of the restricted Poincaré group.

1.3 Comparison

The approaches to the CPT and spin–statistics theorems reviewed in Section 1.2 are summarized in Tables 1.1 and 1.2. Consider, first, how these approaches differ in describing the relation between the properties of CPT invariance and the SSC.

In both the Wightman and algebraic approaches, if a state possesses SSC, then it also possesses CPT invariance, but not vice versa. This is true in the Wightman approach, insofar as the assumptions needed to derive SSC differ from those needed to derive CPT invariance only in the principles of the StLC and WLC, and the former entails the latter.[39] This is also true in the algebraic approach, insofar as the set of irreducible restricted Poincaré-invariant DHR representations with finite statistics and masses (which underwrite states that possess SSC in the algebraic approach) is included in the set of DHR representations (which underwrites states that possess CPT invariance).

In the Lagrangian and Weinberg approaches, it is *not* the case that if a state possesses SSC, then it also possesses CPT invariance. In the Lagrangian approach, while SSC is an explicit assumption in the derivation of CPT invariance, a state may possess SSC but not be CPT invariant, insofar as a state that possesses SSC may be associated with a non-local, Lorentz-invariant, CPT-violating Lagrangian density. In Weinberg's approach, in order for a state that possesses SSC to be CPT invariant, it has to admit a conserved charge. Note, finally, that in both the Lagrangian and Weinberg approaches, CPT invariance of a state entails that state possesses SSC. This holds for the Lagrangian approach, since SSC is a necessary condition for CPT invariance. It also holds for Weinberg's approach, in which the conditions that are sufficient for SSC are necessary for CPT invariance.

These observations are captured in Table 1.3.

Table 1.1 *Alternative formulations of the Spin–Statistics Theorem in RQFTs.*

Approach	Principles	Derived Property
Wightman	(a) Restricted Lorentz invariance (b) Spectrum condition (c) Statistics–locality connection	Spin–statistics connection (SSC) for field states
Algebraic	(a) Modular covariance (b) Additivity (c) Algebraic causality (d) Normal commutation relations	SSC for irreducible, restricted Poincaré-invariant DHR representations with finite statistics and masses
Lagrangian	(a) Restricted Lorentz invariance (b) Spectrum condition (c) Condition (1.1) (d) Causality	SSC for fermionic field states
	(a) Restricted Lorentz invariance (c) Condition (1.1) (d) Causality	SSC for bosonic field states
Weinberg	(a) Restricted Lorentz invariance for S-matrix (b) Cluster decomposition for S-matrix	SSC for particle states

Table 1.2 *Alternative formulations of the CPT Theorem in RQFTs.*

Approach	Principles	Derived Property
Wightman	(a) Restricted Lorentz invariance (b) Spectrum condition (c) Weak local commutativity	CPT invariance for field states
Algebraic	(a) Modular covariance (b) Additivity (c) Algebraic causality	CPT invariance for DHR representations
Lagrangian	(a) Spin–statistics connection (b) Restricted Lorentz invariance (c) Local normal-ordered Hermitian Lagrangian density	CPT invariance of Hamiltonian density for fields
Weinberg	(a) Restricted Lorentz invariance for S-matrix (b) Cluster decomposition for S-matrix (c) Existence of conserved charges	CPT invariance of Hamiltonian density for particles

Table 1.3 *Relations between SSC and CPT invariance.*

Wightman/Algebraic	**Weinberg/Lagrangian**
SSC $\Rightarrow$ CPT invariance	SSC $\nRightarrow$ CPT invariance
CPT invariance $\nRightarrow$ SSC	CPT invariance $\Rightarrow$ SSC

Apart from differing on how they view the relation between CPT invariance and SSC, the approaches also differ in three additional ways. First, RLI is an explicit assumption in all approaches except the algebraic approach, which appeals instead to MC. Second, all the approaches adopt some form of locality constraint: the Wightman approach assumes LC, Weinberg's approach adopts CD, and the Lagrangian and algebraic approaches adopt versions of *causality*. These differing notions of relativity and locality, and the relations among them, are the topic of Chapter 2. Section 1.4 is devoted to a remaining distinction between the approaches that is based, roughly, on how they implicitly treat interactions. This distinction strikes at the heart of a recent debate in the philosophy of quantum field theory.

1.4 Pragmatism versus Purity

Philosophers are split on whether foundational issues related to RQFTs should be framed within one or the other of two basic approaches to formulating RQFTs, what I will call *pragmatism* and *purity*. Pragmatist (or "heuristic") approaches have typically been described as trading mathematical rigor for the ability to derive predictions from realistic interacting theories. Examples include Weinberg's approach and the Lagrangian approach. Purist (or "rigorous") approaches have typically been described as trading the ability to formulate realistic interacting theories for mathematical rigor. Examples include the axiomatic and algebraic approaches. Wallace (2011), for instance, has argued that cutoff quantum field theory (CQFT), a particular pragmatist approach, has been successful at resolving the problems associated with renormalized perturbation theory, while axiomatic and algebraic quantum field theories (AQFTs), which epitomize purist approaches, have not; and this indicates that CQFT is the correct framework for philosophy of QFT. Fraser (2011), on the other hand, argues that renormalization techniques indicate how CQFT and AQFT are empirically indistinguishable, and that AQFT is to be preferred for its mathematical rigor.

Purist approaches like the axiomatic and algebraic formalisms attempt to identify a set of axioms and then construct models of these axioms that describe relevant field theories. These approaches face what may be called the *problem of empirical import*: no "realistic" interacting models of the relevant sets of axioms currently exist. This should be qualified in the following ways.

(i) First, by a *realistic interacting* model, I mean a model for a four-dimensional (4-dim) RQFT from which predictions have been derived and confirmed. Such theories include quantum electrodynamics (QED) that describes the electromagnetic force, the electroweak theory that describes the weak force, and quantum chromodynamics (QCD) that describes the strong force.

(ii) Second, *non-interacting* models for relevant sets of axioms have been constructed.

(iii) Third, non-trivial *unrealistic interacting* models have also been constructed; models for non-trivial interacting theories in two and three dimensions, for instance. In particular, Rivasseau (2003: 168) lists $P(\phi)_2$, ϕ^4_3, the 2- and 3-dim Yukawa model, and the massive 2-dim Gross–Neveu model.

Finally, it should be acknowledged that the fact that no realistic interacting models of purist axioms *currently* exist does not entail that such models cannot *in principle* be constructed. With these qualifications in mind, the *problem of empirical import* suggests that purist understandings of CPT invariance and the SSC (currently) restrict these properties to non-interacting, or unrealistic interacting RQFT states. This is problematic, since the evidence for CPT invariance and the SSC in particular, and for the reliability of RQFTs in general, invariably comes from the successful predictions made by 4-dim realistic interacting RQFTs.

In pragmatist approaches like the Weinberg and Lagrangian formalisms, one might initially be inclined to claim that CPT invariance and the SSC are properties of both non-interacting *and* realistic interacting states. In these approaches, these properties are proven to hold for non-interacting states, but then, typically, an appeal to perturbation theory is taken to justify their ascription to interacting states. One assumes interactions can be expressed in terms of small perturbations about a given state, typically taken to be the vacuum (i.e., the state of lowest energy). One then points out that as long as the interaction Hamiltonian (or Lagrangian) density that encodes these perturbations is a Lorentz scalar, if the free fields (or multi-particle states) that appear in it possess CPT invariance, so will it (provided, in the case of a Lagrangian density, that it is, in addition, Hermitian). Similar remarks can be made about the SSC, insofar as an interacting field (or multi-particle state) is assumed to be expressible in terms of perturbations about a free state. However, lest one think this is an improvement over purist understandings of CPT invariance and the SSC, such pragmatist accounts face the following related problems.

(a) First, in most cases, the goal of pragmatist approaches is to calculate the S-matrix (1.2).[40] The standard method for doing so requires non-interacting multi-particle states at asymptotic times (i.e., at $t = \pm\infty$) to be unitarily related to interacting multi-particle states at finite times. This is made problematic by Haag's theorem, which indicates that, under reasonable assumptions, the Hilbert spaces for interacting and non-interacting states belong to unitarily inequivalent representations of the canonical (anti-) commutation relations, thus no such unitary transformation exists (see, e.g., Earman and Fraser, 2006; Duncan, 2012: 359–370).[41]

(b) A second problem is the fact that, for many realistic interacting QFTs, the power series expansion of the S-matrix contains divergent terms at high energies. This is referred to as the *UV (ultraviolet) problem*.[42]

(c) Finally, for the realistic interacting QFTs of interest, there is a consensus that the power series expansion of the S-matrix does not converge. Call this the *convergence problem*.

These pragmatist problems should be qualified in the following ways:

(i) First, they are common to any approach that employs renormalized perturbation theory to derive predictions from most realistic interacting RQFTs.

(ii) Second, some realistic interacting RQFTs, QCD for instance, may not suffer the *UV problem*. A general consensus understands these theories to possess what is called a UV fixed point (see Section 1.4.3).

(iii) Finally, problem (a) is implicitly addressed in pragmatist approaches by employing renormalization.

In order to further distinguish pragmatists from purists, it will help to review where in pragmatist approaches renormalization occurs.

1.4.1 Pragmatism and the Renormalization Problem

Typical pragmatist approaches simplify equation (1.2) by reducing it to an expression that involves vacuum expectation values of time-ordered products of fields, $\langle 0 | T\{\phi(x_1)\ldots\phi(x_n)\} | 0\rangle$, referred to as τ-functions. One can distinguish between non-interacting and interacting τ-functions, depending on whether the fields are non-interacting or interacting (i.e., satisfy homogeneous or inhomogeneous field equations, respectively). The initial goal of pragmatist approaches is to reduce equation (1.2) to an expression that only involves non-interacting τ-functions.[43] In practice, this goal is achieved by appealing to the following two results:

(i) One first makes use of a non-perturbative result due to Lehmann, Symmanzik, and Zimmermann (1955) that allows S-matrix elements to be calculated from interacting τ-functions (see, e.g., Duncan, 2012: 286). This is known as the LSZ reduction formula and comes in many flavors, one per type of field. For instance, the LSZ formula for a scalar field of mass m is given by:

$$
\begin{aligned}
&{}_{out}\langle \mathbf{p}_1, \ldots, \mathbf{p}_n | \mathbf{k}_1, \ldots, \mathbf{k}_\ell \rangle_{in} \\
&= \left(i/\sqrt{Z}\right)^{n+\ell} \int d^4x_1 \ldots d^4y_\ell e^{-ip_ix_i+ik_jy_j} \prod_i \left(\partial^2_{x_i} + m^2\right) \prod_j \left(\partial^2_{y_j} + m^2\right) \\
&\quad \times \langle 0 | T\{\varphi(x_1)\ldots\varphi(x_n)\varphi(y_1)\ldots\varphi(y_\ell)\} | 0 \rangle.
\end{aligned} \tag{1.3}
$$

The left-hand side of equation (1.3) represents an S-matrix element for ℓ incoming particles with momenta k_i and n outgoing particles with momenta p_i. The right-hand side indicates how this can be calculated in terms of an interacting τ-function, where $\varphi(x)$ is an interacting field (i.e., a solution to the inhomogeneous Klein–Gordon equation).

(ii) One then assumes a perturbative split of the Hamiltonian, $H = H_0 + H_{int}$, into a non-perturbed piece H_0 (typically identified as the free Hamiltonian) and a piece H_{int} encoding small perturbations away from H_0.[44] The following formula due to Gell-Mann and Low (1951) then allows interacting τ-functions to be calculated from non-interacting τ-functions (see, e.g., Duncan, 2012: 246):

$$
\langle 0 | T\{\varphi(x_1)\ldots\varphi(x_n)\} | 0 \rangle = \frac{\langle 0 | T\{\phi_I(x_1)\ldots\phi_I(x_n)e^{-i\int H_I dt}\} | 0 \rangle}{\langle 0 | T\{e^{-i\int H_I dt}\} | 0 \rangle}. \tag{1.4}
$$

In equation (1.4), $\varphi(x)$ is an interacting field, $\phi_I(x)$ is a non-interacting field in the interaction picture, and $H_I \equiv e^{iH_0} H_{int} e^{-iH_0}$ is the interaction picture representation of H_{int}.[45] The left-hand side of equation (1.4) represents an interacting τ-function that enters into the LSZ formula, and the right-hand side indicates how it can be calculated in terms of free τ-functions.

In the LSZ formula (1.3), the constant Z is a renormalization constant. Its purpose is to relate the interacting field $\varphi(x)$ to non-interacting fields $\phi_{in}(x)$, $\phi_{out}(x)$ at asymptotic times. One assumes,

$$
\langle \beta | \varphi(x) | \alpha \rangle \xrightarrow[t\to-\infty]{} \sqrt{Z} \langle \beta | \phi_{in}(x) | \alpha \rangle, \; \langle \beta | \varphi(x) | \alpha \rangle \xrightarrow[t\to+\infty]{} \sqrt{Z} \langle \beta | \phi_{out}(x) | \alpha \rangle \tag{1.5}
$$

where $|\beta\rangle$, $|\alpha\rangle$ are non-interacting multi-particle states. Under this weak notion of convergence, matrix elements of interacting fields are required to converge to matrix elements of non-interacting, asymptotic fields at asymptotic times (as opposed to requiring that interacting fields converge directly to non-interacting fields). This assumption may be motivated by considering the action of a non-interacting asymptotic field on the vacuum with respect to a single-particle state (Duncan, 2012: 282). If $|\mathbf{k}\rangle$ is a normalized non-interacting single-particle state, then $\langle\mathbf{k}|\phi_{in}(x)|0\rangle = 1$. An interacting field $\varphi(x)$ cannot, in general, be decomposed into non-interacting creation and annihilation operators, thus one sets $\langle\mathbf{k}|\varphi(x)|0\rangle = \sqrt{Z}$, for some constant Z. Equation (1.5) may be considered a generalization of this. Formally, the constant Z can be removed from the LSZ formula by replacing the "bare" interacting field with a renormalized interacting field defined by $\varphi_r(x) \equiv Z^{-1/2}\varphi(x)$. This assignment guarantees that the renormalized interacting field behaves like the non-interacting field with respect to single-particle states; namely, $\langle\mathbf{k}|\varphi_r(0)|0\rangle = 1$.

Renormalization also enters into the derivation of the Gell-Mann–Low formula (1.4). In particular, formula (1.4) assumes that the non-perturbed Hamiltonian and the full Hamiltonian both annihilate the vacuum: $H_0|0\rangle = 0 = H|0\rangle$. The first equality entails $|0\rangle$ is the vacuum state of the non-interacting fields. Since H is a functional of interacting fields which, again, cannot in general be decomposed into non-interacting creation and annihilation operators, the second equality is typically not guaranteed. To enforce it, one defines a renormalized Hamiltonian $H_r \equiv H - \Delta$. This corresponds to renormalizing the mass that appears in H. If this is given by m_B (the "bare" mass), and the shift corresponding to Δ is given by δm, then the renormalized mass m_r (the "physical" mass) is given by ${m_r}^2 \equiv {m_B}^2 + \delta m^2$.

In these examples, renormalization is imposed to force the interacting theory to behave like the non-interacting theory, as far as the vacuum and single-particle states are concerned.[46] This solves the pragmatist's problem (a) in the following sense: the renormalized field and the renormalized Hamiltonian are not self-adjoint operators (for typical realistic interacting theories, the constant Z and the mass shift δm are infinite). This entails, for instance, that H_r does not implement unitary time translations, contrary to one of the assumptions of Haag's theorem (Fraser, 2009: 547).[47] Whether this constitutes an *adequate* solution to Problem (a) will depend on one's mathematical proclivities. The fact that renormalized parameters are, typically, infinite may upset purists. For such purists, the pragmatist's renormalization procedure simply replaces problem (a) with another problem, call it the *renormalization problem*.[48]

Note that renormalization is independent of perturbation theory insofar as renormalization appears already in the non-perturbative derivation of the LSZ formula (as well as in the non-perturbative Källen–Lehman representation of the interacting Feynman propagator mentioned in footnote 46). Perturbation theory enters explicitly into the derivation of the Gell-Mann–Low formula, which in practice is calculated by expanding the exponentials on the right-hand side of equation (1.4) as power series in the coupling constants that appear in H_I. The *UV* and *convergence problems* then re-emerge as problems associated with these power series. Thus the *renormalization problem* is independent of the *UV* and *convergence problems*.[49]

However, there is a formal sense in which renormalization solves the *UV problem*. The types of divergences associated with perturbative expansions of interacting τ-functions can in principle be regularized by either renormalizing the parameters that are present in the Lagrangian density, or by introducing new renormalized parameters. For example, in ϕ^4 theory (i.e., massive scalar field theory with a quartic interaction term $\lambda\phi^4$), the divergences that appear in the perturbative expansion of the 2-point τ-function can be regulated by renormalizing the mass and the field, whereas the divergences that appear in the perturbative expansion of the 4-point τ-function can be regulated by renormalizing the coupling constant λ.[50] In principle, all divergences in higher-order τ-functions can also be regularized by introducing new terms into the Lagrangian density and renormalizing the parameters associated with them. The 6-point τ-function, for instance, can be regularized by introducing a new term $\lambda_{(6)}\phi^6$ into the Lagrangian density and renormalizing the parameter $\lambda_{(6)}$. In general, if this process never ends, i.e., if for each higher-order τ-function one needs to introduce new terms in the Lagrangian density in order to regularize it, then the theory is referred to as *non-renormalizable*. On the other hand, if only a finite number of renormalized parameters are needed to regularize all possible divergent terms in the perturbative expansions of τ-functions, then the theory is referred to as *renormalizable*. Thus, formally, renormalization solves the *UV problem*, for both non-renormalizable and renormalizable theories. However, since the solution to the problem for non-renormalizable theories requires an infinite number of renormalized parameters, and since the values of these parameters are fixed by experiments, a non-renormalizable theory has no predictive power, all things being equal. Thus, without further ado, renormalization only solves the *UV problem* for renormalizable theories.[51]

One can make further ado by inserting a UV cutoff and then demonstrating that, within the energy range specified by the cutoff, only a finite number of terms need to be renormalized in order to derive predictions from the theory. This further ado forms the basis of the renormalization group (RG) approach to renormalization. This approach solves both the *UV* and *renormalization problems* for the pragmatist, but typically the *convergence problem* remains. Among philosophers, Wallace (2011) has argued that RG techniques underwrite heuristic (i.e., pragmatist) approaches, whereas Fraser (2011) claims they support rigorous (i.e., purist) approaches. The next section addresses this issue, as well as the general concern of how best to distinguish pragmatism from purity.

1.4.2 Distinguishing Pragmatism from Purity

The goal of the RG approach to renormalization is to determine how a theory's low-energy degrees of freedom depend on its high-energy degrees of freedom. Toward this end, the coupling constants g that appear in the interaction Hamiltonian (or Lagrangian) density, are defined as functions $g(\Lambda(\mu))$ of a scale-dependent cutoff $\Lambda(\mu)$. Changing the scale (by integrating out high-energy degrees of freedom with respect to Λ) generates a flow in the theory's parameter space.

This flow is characterized by the beta-function, defined by (see, e.g., Duncan, 2012: 581):

$$\beta(g) = \mu \frac{dg}{d\mu} \tag{1.6}$$

Couplings can be characterized by how they behave as the scale is lowered: relevant couplings increase, irrelevant couplings decrease, and marginal couplings remain constant. One can then show that, for a (3+1)-dim weakly coupled theory, there are a finite number of relevant and marginal couplings, and any irrelevant couplings are suppressed at a given energy scale μ by powers of μ/Λ.[52] In such a theory, the low-energy degrees of freedom depend on the high-energy degrees of freedom through a finite number of parameters (the relevant and marginal couplings), and while the theory may still contain parameters that become infinite at high energies (the irrelevant couplings), it is still predictive in the sense that its predictions will be finite if constrained to a given scale. At this scale, the theory is *effectively* renormalizable insofar as any irrelevant couplings it may possess cannot be experimentally detected. With respect to the discussion of renormalization in Section 1.4.1, an effectively renormalizable interacting theory requires a finite number of parameters to empirically imitate the behavior of the corresponding non-interacting theory, and these parameters, as functions of a finite cutoff, are themselves finite.

In this effective field theory approach, the *renormalization problem* is addressed by adopting effective renormalizability, and the *UV problem* is addressed by using the cutoff Λ to regulate divergent terms in expressions like equation (1.2). The cutoff serves to freeze out the high energy degrees of freedom of the theory, and one then adopts an agnostic attitude about what happens at energy scales above Λ. According to Wallace,

> This, in essence, is how modern particle physics deals with the renormalization problem: it is taken to presage an ultimate failure of quantum field theory at some short length scale, and once the bare existence of that failure is appreciated, the whole of renormalization theory becomes unproblematic, and indeed predictively powerful in its own right. (Wallace, 2011: 119)

While this appeal to RG techniques allows a pragmatist to address the *renormalization* and *UV problems*, the *convergence problem* still remains. Moreover, Fraser (2011) suggests that RG techniques support purity, as opposed to pragmatism. In particular, the RG flow of the type of theory described above indicates an underdetermination of the theory's high-energy content by low-energy experiments. The latter fix the values of the theory's finite relevant and marginal couplings at the experimental energy scale, but fail to fix the values of the theory's irrelevant couplings, and it is these which determine how the theory behaves at high energies. This implies that the successful predictions made by a realistic interacting RQFT (of this type) fail to determine the form it takes at high energies. This suggests to Fraser that axiomatic and algebraic RQFT (AQFT), on the one hand, and

Wallace's "cutoff" QFT (CQFT), on the other, are empirically indistinguishable at the energy scales currently probed by experiments:

> The upshot of the application of RG methods is that a range of Lagrangians at short distance scales each yield approximately the same predictions for relatively low energies... This lends support to the claim that the theoretical framework of QFT is underdetermined by the empirical evidence. AQFT and [CQFT] should be viewed as alternative theoretical frameworks for QFT which approximately agree in their empirical predictions. (Naturally, subject to the qualification that the construction of models of AQFT is still in progress). (Fraser, 2011: 135)

The qualification at the end of this quote is important. It acknowledges that the purist's *problem of empirical import* is a potential obstruction to the claim that RG underdetermination holds between AQFT and CQFT. This obstruction takes the form of the question of whether there are AQFTs that can be "RG-related" to the appropriate low-energy experiments (as CQFTs can be).

These considerations suggest that an appeal to RG techniques is not decisive in adjudicating between pragmatists and purists. Both pragmatists and purists can make such an appeal, and these appeals fail to completely address foundational issues: the *convergence problem* remains for the RG pragmatist and the *problem of empirical import* remains for the RG purist. Note, too, that an appeal to perturbation theory won't help either. On the one hand, pragmatists can employ non-perturbative techniques (the LSZ formula, for example; and lattice techniques in theories like QCD that are not weakly coupled). On the other hand, purists can employ perturbative techniques, as evidenced by "perturbative" AQFT which seeks to combine techniques from causal perturbation theory with the algebraic formalism.[53] The diversity of such methods allowed by both pragmatists and purists also suggests that a general appeal to mathematical rigor may not be enough to make the distinction as clear as it could be.

1.4.3 The Existence Problem

What remains to distinguish pragmatists from purists are the *convergence problem* for pragmatists and the *problem of empirical import* for purists. These problems are concerned with the sense in which realistic interacting RQFTs can be said to exist. Call this basic foundational concern common to both purity and pragmatism, the *existence problem*. As Bouatta and Butterfield (2015: 68–69) suggest, it can be addressed in a number of distinct ways.

Purists seem to have a fairly straight-forward notion of existence; namely, existence of a model of an appropriate set of axioms. Recall that this criterion is made problematic by the fact that models describing realistic (viz., 4-dim) interacting RQFTs have yet to be constructed.

With respect to existence, pragmatists have at least three distinct options, all arguably weaker than the purist criterion.

(i) A pragmatist might require the existence of a theory to entail the convergence of perturbative series expansions of relevant quantities like equation (1.2). Under this criterion, weakly interacting theories like QED and electro-weak theory do not exist, since the perturbative expansions associated with these theories are divergent asymptotic series.[54] Such series have the property that successive terms decrease in magnitude only for some finite number of orders before starting to increase again (Duncan, 2012: 386). Strongly interacting theories like QCD also do not exist under this criterion. In a strongly interacting theory, one can still make use of perturbation theory to calculate relevant quantities, but now at high (as opposed to low) energies. That this can be done reliably for QCD is secured by the claim that QCD is asymptotically free: the strong interaction tends to zero at high energies (see criterion (iii) below). However, again, such perturbative series for relevant quantities in QCD are asymptotic. Note that one can adopt a weaker perturbative criterion of existence than convergence. For instance, one might require that the perturbative series be "Borel summable" (Weinberg, 1996: 283; Duncan, 2012: 400). A Borel summable divergent asymptotic series can be reconstructed, in principle, from a convergent series obtained by performing a Borel transformation on the original series.[55] However, Duncan (2012: 403) indicates that "... the property of Borel summability is an extremely fragile one, and one which we can hardly ever expect to be present in interesting relativistic field theories." In particular, the original series typically must be renormalized (recall this requires replacing the bare couplings with cutoff independent renormalized couplings). This process of renormalization introduces singularities (referred to as "renormalons") in the Borel transformed series, and these prevent one from summing the series to obtain an exact magnitude. Thus even under the weaker existence criterion of Borel summability, arguably, realistic interacting RQFTs do not exist.

(ii) Alternatively, a pragmatist might settle for existence to be defined in terms of renormalizability. Again, a renormalizable theory is one in which the couplings depend on a cutoff in such a way that each term is finite and cutoff independent. Under this criterion, interacting QED, interacting electroweak theory, and interacting QCD all exist. On the other hand, the current consensus among pragmatists seems to be that renormalizability is too strong a criterion to impose on a theory, at least as a criterion of acceptance. This consensus claims that non-renormalizable theories can be used effectively in deriving and confirming predictions.[56] Again, this effective field theoretic approach takes an agnostic attitude toward what goes on at high energies, and this allows peaceful coexistence between relevant and marginal (viz., renormalizable/super-renormalizable) couplings on the one hand, and irrelevant (viz., non-renormalizable) couplings on the other. Under this attitude, for instance, one can safely use the quantum field theoretic formulation of general relativity to derive predictions at low energy scales, even though this theory contains an infinite number of irrelevant couplings.

(iii) A final example of a pragmatist existence criterion is the existence of a UV fixed point in a RG flow. A UV fixed point is a finite limit that the coupling constants g tend to as the energy scale μ tends to infinity. In terms of the beta-function (1.6), a UV fixed point is characterized by $\lim_{\mu\to\infty} \beta = 0$. Intuitively, theories that possess UV fixed points do not blow up at high energies. Under this criterion, interacting QED and interacting electroweak theory do not exist (these theories do not possess a UV fixed point), whereas the consensus is that interacting QCD does. Bouatta and Butterfield (2014: 27) distinguish three types of theories that satisfy this criterion, based on the following properties:

   (a) Asymptotic freedom: $\lim_{\mu\to\infty} \beta = \lim_{\mu\to\infty} g = 0$.

   (b) Asymptotic safety: $\lim_{\mu\to\infty} \beta = 0$; $\lim_{\mu\to\infty} g = g_* \neq 0$.

   (c) Conformal invariance: $\beta = 0$.

   An interacting *asymptotically free* theory possesses a UV fixed point that represents the non-interacting theory (referred to as a Gaussian fixed point), i.e., as the energy scale tends to infinity, the theory's couplings tend to zero. Interacting QCD displays this type of behavior. An interacting *asymptotically safe* theory possesses a UV fixed point that is not necessarily Gaussian. Weinberg (1979) suggested that the formulation of general relativity as a quantum field theory might behave in this way (this has spawned a research program that attempts to identify UV fixed points of general relativity; see, e.g., Bain 2014). Finally, a *conformally invariant* theory trivially possesses a UV fixed point, in the sense that its couplings remain constant at all energy scales. Conformally invariant theories play important roles in string theory and twistor theory; however, none of the realistic interacting RQFTs of interest possess this property.

The above considerations provide a mixed bag when it comes to assessing the debate between purists and pragmatists. On the one hand, distinguishing purity from pragmatism on the basis of the *existence problem* addresses the fact that both purists and pragmatists make use of similar methods, perturbative and non-perturbative, as well as the concern that mathematical rigor may be in the eye of the beholder. On the other hand, neither approach can currently provide a complete solution to this problem.

1.5 Summary

Why are CPT invariance and the SSC derivable properties in RQFTs? We have seen that there are at least four ways to answer this question that, at least on the surface, are underwritten by distinct commitments to what RQFTs are about and what basic principles they are required to satisfy. Taken at face value, the Wightman and Lagrangian approaches view RQFTs to be fundamentally about fields, yet they differ over the principles such fields must satisfy in order to derive CPT invariance and the SSC. Weinberg's approach views RQFTs to be about particles governed by the *S*-matrix, and subsequently identifies constraints that the *S*-matrix must satisfy in order to derive CPT invariance and the SSC. Finally, in the context of the CPT and spin–statistics theorems, the algebraic approach commits itself to a net of local observable algebras realized as operators on a Hilbert space, and adopts yet another set of basic principles to underwrite CPT invariance and the SSC. As Section 1.3 indicated, relativity, in the guise of RLI, is not a common feature of all approaches; in particular, the algebraic approach appeals, instead, to MC. Moreover, all approaches include additional principles, including various versions of a locality constraint. These versions, and their relation to RLI and MC, will be investigated in the first part of Chapter 2. At this stage, however, it's safe to say that the reason why CPT invariance and the SSC are derivable properties in RQFTs is not *solely* due to relativity. In fact, one can argue that relativity is neither necessary (algebraic approach), nor sufficient to derive CPT invariance and the SSC, appropriately construed. This should give us pause in thinking that relativity is the answer to the related question, to be considered in Chapters 3 and 4, "Why are CPT invariance and the SSC not derivable properties in non-relativistic quantum theories?"

The proofs of the CPT and spin–statistics theorems in RQFTs reviewed in Section 1.2 can be sorted into two basic types: purist and pragmatist. Both types demonstrate that CPT invariance and the SSC are essential properties of *non-interacting* and at most *unrealistic interacting* RQFTs. Any extension of these theorems to *realistic interacting* RQFTs (and in particular, to the 4-dim RQFTs that constitute the standard model) must confront the *existence problem* of Section 1.4.3. The task for purists is to construct a realistic interacting model of an appropriate set of axioms. The task for pragmatists is to demonstrate that their preferred notion of existence holds for the realistic interacting theories of interest. This distinction between purity and pragmatism will be put to work in the second part of Chapter 2 where it will be used to assess an influential yet puzzling claim made by Greenberg (2002) about the relation between CPT invariance and Lorentz invariance in interacting RQFTs.

NOTES

1. Wallace (2009) demonstrates that whether or not there are matter/antimatter representations of a symmetry group depends on whether the complexification of the group is irreducible. If it isn't, then it always admits two irreducible representations related by charge conjugation.

2. See Baker (2009) and Wallace (2006) for reviews of Fock space and wavefunctional space.
3. As Baker (2009: 606) points out. Streater and Wightman's (1964: 100–101) exposition of the Wightman approach requires that "To be a field theory, a relativistic quantum theory must have enough fields so its states can be uniquely characterized using fields and functions of fields." They view a sufficient but not necessary condition for this to be that the fields satisfy the equal times commutation relations, which is a necessary ingredient in the construction of a wavefunctional representation. They suggest a weaker condition; namely, that there exists a unique vacuum state which is cyclic in the fields (i.e., the set of states generated by operating on the vacuum with any field is dense in the Hilbert space of states).
4. See, e.g., Weinberg (1995: 74–76, 121) and Roberts (2014). Following Weinberg, one can show that P and T stand in the following relations with the generator H of time translations in the Lorentz Lie algebra: $\mathsf{P}iH\mathsf{P}^{-1} = iH$, $\mathsf{T}iH\mathsf{T}^{-1} = -iH$. Under the assumptions that P is antiunitary and T is unitary, one would have $\mathsf{P}H\mathsf{P}^{-1} = -H$ and $\mathsf{T}H\mathsf{T}^{-1} = -H$. This would entail that for any state $|\Psi\rangle$ with positive energy, there would be states $\mathsf{P}^{-1}|\Psi\rangle$, $\mathsf{T}^{-1}|\Psi\rangle$ with negative energy, which violates the SC.
5. Suppose $|\Phi\rangle$ is a multi-particle state, and let $|\Phi'\rangle$ be a multi-particle state obtained from $|\Phi\rangle$ by permuting its single-particle substates. $|\Phi\rangle$ is *symmetric* just when $|\Phi'\rangle = |\Phi\rangle$. $|\Phi\rangle$ is *antisymmetric* just when $|\Phi'\rangle = -|\Phi\rangle$. $|\Phi\rangle$ is *permutation invariant* just when, for any linear operator A representing an observable quantity, the expectation value of A is the same for $|\Phi\rangle$ and $|\Phi'\rangle$: $\langle\Phi|A|\Phi\rangle = \langle\Phi'|A|\Phi'\rangle$.
6. Haag (1996: 97) takes the first route. Streater and Wightman (1964: 147) suggest the second route: "A natural way to arrive at Bose–Einstein statistics is to describe the system in question by a field which commutes for spacelike separations, while the analogous way for Fermi–Dirac statistics is to use a field which anticommutes for spacelike separations." The "natural way," evidently, would be to demonstrate that StLC entails that, when Fock space creation/annihilation operators corresponding to the fields exist, they satisfy equation (1.1).
7. This relation is motivated by Greenberg's (1998) insightful distinction between the "spin–statistics" theorem and the "spin–locality" theorem, which is discussed in Section 2.1.4 of Chapter 2.
8. A group G' is a double-cover of another group G just when there is a group homomorphism $\rho: G' \rightarrow G$ such that to every element of G' there corresponds exactly two elements of G.
9. The Wightman axioms are reproduced in Section 3.2.1. Expositions can be found in Streater and Wightman (1964), Haag (1996), and Araki (1999).
10. This requires an appeal to the "separating corollary" (Theorem 4-3 in Streater and Wightman, 1964: 139). See the Appendix for a discussion. The complete derivation is given in Streater and Wightman (1964: 148–150).

11. Why should we assume the only statistics in the running are Bose–Einstein (BE) or Fermi–Dirac (FD)? As footnote 4 of the Introduction indicates, there is good reason to believe that any theory that describes non-standard statistics (i.e., "parastatistics") is a notational variant of an adequately formulated theory of either BE or FD statistics.
12. A Jost point $(x_1, \ldots, x_n)$ is a convex set of real points that are spacelike separated from each other. In other words, the difference variables $\xi_i = x_{i-1} - x_i$ satisfy $(\sum \lambda_j \xi_j)^2 < 0$, for $\lambda_j \geq 0$, $\sum \lambda_j > 0$ (Streater and Wightman, 1964: 71). One can show that the extended domain on which complex Wightman functions are defined contains Jost points.
13. This follows Haag (1996: 98) and Greenberg (2006b: 1547). Streater and Wightman (1964: 131, 145) identify F with the number of half-integer-spin fields.
14. One can show that if a complex Wightman function possesses a property (like WLC) at a Jost point, then it possesses that property at all (complex) points in the extended domain. It is then easy to show that if complex Wightman functions possess both WLC and PT, then they obey a "CPT condition." Taking the boundary limit $\eta_i \to 0$ of this CPT condition for complex Wightman functions produces a CPT condition for real Wightman functions, which entails the existence of a CPT operator on fields.
15. One must also assume that the fields have finitely many components to avoid counterexamples of infinite fields that do not possess the SSC or CPT invariance (Streater, 1967; Oksak and Todorov, 1968).
16. Massimi and Redhead (2003) compare Weinberg's approach to the spin–statistics theorem with the standard textbook approach reviewed in Section 1.2.3 below.
17. For a derivation of equation (1.2), and a discussion of the significance of time-ordered products, see Duncan (2012: 70–75) and Weinberg (1995: 141–145).
18. This commutativity condition guarantees that time-ordered products of $\mathfrak{H}_{int}(x)$ are restricted Lorentz invariant. See Section 2.1.1 for further explanation.
19. More precisely, a sufficient condition for (iii′) is that the interaction Hamiltonian density takes the form of a sum of creation and annihilation operators with coefficient functions that contain just a single 3-dim momentum conservation delta function. Section 2.1.3 provides an outline of the argument given by Weinberg (1995: 182), who claims ("as far as I know") that this condition is both sufficient *and* necessary.
20. Again, $\phi(x)$ is supposed to represent a generic field with arbitrary spin. Weinberg's notation makes use of indices to represent this fact; these indices are suppressed in the following exposition.
21. That property (α) entails CD is explained in Section 2.1.3. Property (β) guarantees that $\mathfrak{H}_{int}(x)$ is RLI, whereas property (γ) guarantees that it commutes at spacelike separated distances "... provided that [it] contains an even

number of fermion field factors" (Weinberg, 1964: 1318). This quote suggests that Weinberg has StLC in mind for property (γ); however, the ensuing argument indicates that it is really LC.

22. This is reflected in Weinberg's (1995: 198) view of LC: "The point taken here is that [LC] is needed for the Lorentz invariance of the S-matrix, without any ancillary assumptions about measurability or causality."
23. The existence of a conserved charge entails that $\mathfrak{H}_{int}(x)$ must commute with the charge operator Q. This entails that $\mathfrak{H}_{int}(x)$ must be formed out of fields $\phi(x)$ that have simple commutation relations with Q. To accomplish this, it suffices to construct $\phi(x)$ as a sum $\phi(x) = \phi^{+}(x) + \phi^{+c\dagger}(x)$, where $\phi^{+}(x)$ and $\phi^{+c}(x)$ are linear combinations of creation/annihilation operators a, a^c for particle states with the same mass and spin, but opposite charge, i.e., $\phi(x)$ is a sum of fields associated with particles and their antiparticles.
24. One can bypass this choice from the start by imposing StLC on the fields, as opposed to LC, and defining bosonic versus fermionic fields accordingly, with no reference to particle states.
25. This interpretation of observables is also expressed by Peskin and Schroeder (1995: 56), and is implicit in Pauli's (1940: 721) original argument.
26. This entailment can be compared with Weinberg's entailment (b) in Section 1.2.2, which can be put into the form [Condition (1.1) & ~SSC] $\Leftrightarrow$ ~LC. Whereas one can argue that ~SSC is equivalent to NSSC (see Section 1.2.1), it is not the case that ~LC is equivalent to NSpLC. (Note, too, that neither is ~SpLC equivalent to NSpLC, i.e., the *failure* of the spin–locality connection is not the same as the *wrong* spin–locality connection, insofar as the failure of a commutator (resp. anticommutator) to vanish does not imply the corresponding anticommutator (resp. commutator) vanishes.)
27. The violation of *causality* in these arguments implicitly assumes the observables of a theory are integer-spin fields or bilinears in half-integer-spin fields. For instance, the standard example of (b) involves the demonstration that if a neutral spin-0 scalar field is quantized via anticommutation relations imposed on its creation and annihilation operators, then the commutator of the field variables is given by a function Δ_1 that does not vanish at spacelike separations.
28. See also Bjorken and Drell (1965: 124), Rebenko (2012: 246), and Sozzi (2008: 197).
29. One motivation for this requirement is that the product of two fields at a point is typically undefined, and normal-ordering is a method of regularizing the resulting divergences.
30. According to Kaku (1993: 122), a non-local theory contains terms of the generic form $\phi(x)\phi(y)$, and these can be power expanded as $\phi(x)\phi(y) = \phi(x)e^{(y-x)^{\mu}\partial_{\mu}}\phi(y)$, which contains, in principle, an infinite number of derivative terms. Note also that non-local Lorentz invariant Lagrangian densities can be constructed that violate CPT invariance (Chaichian et al., 2011).

31. For expositions of this approach, see, e.g., Araki (1999), Haag (1996), and Halvorson and Müger (2006). The Haag–Araki axioms on which it is based are reproduced in Section 3.2.4.
32. Ω is cyclic for $\mathfrak{R}$ just when $\{A\Omega : A \in \mathfrak{R}\}$ is dense in $\mathcal{H}_0$.
33. A wedge region in Minkowski spacetime M is any Poincaré transformation of the region$\{x \in M : x_1 > |x_0|\}$, where (x_0, x_1, x_2, x_3) is an inertial coordinate system. That a modular operator exists is entailed by the Tomita–Takesaki theorem (Haag, 1996: 217; Halvorson and Müger, 2007: 738). This theorem demonstrates that a von Neumann algebra $\mathfrak{R}$ of bounded linear operators on a Hilbert space $\mathcal{H}$ with a cyclic and separating vacuum vector Ω possesses a one-parameter collection of modular operators $\Delta(t)$ and a modular conjugate operator $\mathcal{J}$ such that $\mathcal{J}\Omega = \Omega = \Delta\Omega$, $\Delta^{it}\mathfrak{R}\Delta^{-it} = \mathfrak{R}$ and $\mathcal{J}\mathfrak{R}\mathcal{J} = \mathfrak{R}'$, where $\mathfrak{R}'$ is the commutant of $\mathfrak{R}$ (i.e., the set of bounded linear operators that commute with all elements of $\mathfrak{R}$).
34. For an outline of the role of the field algebra in the algebraic approach to superselection theory, see Halvorson and Müger (2007: Section 9).
35. In general, a representation of $\mathfrak{R}$ consists of a pair $(\mathcal{H}, \pi)$ where $\mathcal{H}$ is a Hilbert space and π is a map that takes elements of $\mathfrak{R}$ to bounded linear operators on $\mathcal{H}$. A state ω on $\mathfrak{R}$ is a linear map that takes elements of $\mathfrak{R}$ to complex numbers. The GNS theorem entails that any state can be associated with a unique (up to unitary equivalence) representation (Araki, 1999: 34; Halvorson and Müger, 2007: 734).
36. Both of these representations have dimension 1. Higher dimension representations of S_N correspond to parastatistics.
37. $\Theta = \mathcal{J}_W R_W$, where $\mathcal{J}_W$ is the modular conjugate operator of $\mathfrak{R}$ restricted to the wedge W, and R_W implements rotations that leave W invariant.
38. Doplicher et al. (1974) derived the SSC for irreducible, restricted Poincaré-covariant DHR representations with finite statistics, positive masses, and finitely many components, under the assumptions of *algebraic causality, Haag duality*, and *property B* (for definitions of the latter, see Araki, 1999: 163–164, or Halvorson and Müger, 2007: 784). Guido and Longo's approach recovers this result in the following way: first, *algebraic causality, additivity*, and MC entail *essential duality*, which is a weaker form of *Haag duality* that still allows Doplicher, Haag, and Robert's analysis to go through. Second, the existence of a unique unitary representation $U\rho$ of the restricted Poincaré group that satisfies the SC has two consequences: (a) uniqueness rules out counterexamples to the spin–statistics theorem of fields with infinitely many components (Guido and Longo, 1995: 519); and (b) *algebraic causality*, SC, and a weaker version of *additivity* entail *property B* (Halvorson and Müger, 2007: 748).
39. The statistics–locality connection (StLC) entails WLC *provided* the integer F that appears in WLC is identified as the number of fermionic fields. See footnote 13.
40. In some important cases, the fundamentality of the S-matrix might be questioned. In asymptotically free theories like QCD in which interactions increase

at low energies, there are no non-interacting asymptotic multi-particle states associated with the fields (quark and gluon) that appear in the Lagrangian density. Since an S-matrix requires the existence of non-interacting asymptotic states, one might conclude that a QCD S-matrix cannot be constructed. As Dütsch and Gracia-Bondía (2012: 432) observe, "One has to admit that large parts of the Standard Model disown the basic hypothesis of any S-matrix theory." On the other hand, asymptotic states can be associated with products of quark and gluon fields (in the sense that such products can define interpolating fields for asymptotic single-particle states). Thus, according to Duncan (2012: 301), "In such confining theories we may loosely speak of the physical particle states as (stable or unstable) bound states of the underlying quark and gluon 'particles', even though there is no attainable physical circumstance in which these latter can be realized as isolated entities with the characteristics expected of a particle..."

41. The "standard method" referred to in the text that falls victim to Haag's theorem is the interaction picture, which assumes the existence of a unitary transformation between interacting and non-interacting states. Haag's theorem (in one of its versions) is based on the following assumptions: for two local fields ϕ_1, ϕ_2, (a) the fields belong to irreducible representations of the equal-time canonical commutation relations, (b) the fields possess unique Euclidean-invariant vacuum states, (c) there is a unitary transformation that relates the fields at a given time, and (d) the fields and vacuum states are Poincaré-invariant. One can then show that if one of the fields is free, both must be free (Earman and Fraser, 2006: 314). The "reasonable assumptions" referred to in the text are assumptions (a), (b), and (d). Assumptions (a) and (b) seem reasonable since they are necessary for the construction of a Fock space, whereas assumption (d) is, of course, necessary in RQFTs. Thus if we adopt assumptions (a), (b), and (d), then assumption (c) must be denied.
42. This problem can also occur at low-energies, in which case it is called the *IR (infrared) problem*. The latter is primarily of concern for theories with massless interactions, like QED.
43. Once this goal has been achieved, standard pragmatist approaches then make use of Wick's theorem to reduce a free n-point τ-function to a sum of products of free 2-point τ-functions. The latter define Feynman propagators, and one can then employ the machinery of Feynman diagrams to calculate equation (1.2).
44. The Hamiltonian is related to the Hamiltonian density by $H(t) = \int d^3\mathbf{x}\,\mathfrak{H}(\mathbf{x}, t)$.
45. $\phi_I(x)$ is defined by $\phi_I(\mathbf{x}, t) \equiv e^{iH_0(t-t_0)}\phi(\mathbf{x}, t_0)e^{-iH_0(t-t_0)}$, where $\phi(\mathbf{x}, t_0)$ is a non-interacting field at time t_0.
46. Another way to see this is by an analysis of the non-perturbative Källen–Lehman representation of the interacting 2-point τ-function (Weinberg, 1995: 439–49).

47. This is the assumption that the fields and vacuum states are Poincaré-invariant, which requires that Poincaré transformations, including time translation, be represented by unitary operators.
48. Duncan (2012: 370) and Wallace (2006: 58) address Haag's theorem by adopting an IR (or long-distance) cutoff. This violates Euclidean invariance of the vacuum state, which is one of the assumptions underlying (one version of) Haag's theorem (see footnote 41). This is the route of Wallace's (2011) "cutoff" QFT, discussed in Section 1.4.2.
49. As Weinberg (1995: 441) states, "... the renormalization of masses and fields has nothing directly to do with the presence of infinities, and would be necessary even in a theory in which all momentum space integrals were convergent."
50. This description of renormalization in ϕ^4 theory is given in Maggiorie (2005: 135).
51. Moreover, this renormalization solution to the *UV problem* may exacerbate the *convergence problem* with the introduction of UV (and/or IR) "renormalons" (Rivesseau, 1991: 6; Duncan, 2012: 403).
52. The proof is based on linear stability arguments and originally given in Polchinski (1984). Duncan (2012: 654–656) provides a summary. An important exception to the proof is QCD, which is not weakly coupled.
53. See, e.g., the review by Summers (2012: 45–48).
54. The series $F(\lambda) = \sum_n c_n \lambda^n$ is said to be asymptotic to a function $f(\lambda)$ if $|f(\lambda) - \sum_{n=0}^{N} c_n \lambda^n| = O(\lambda^{N+1})$ as $\lambda \to 0$ for any fixed N (Duncan, 2012: 377).
55. See, e.g., Weinberg (1996: 283). The Borel transform of the series $F(\lambda) = \sum_n c_n \lambda^n$ is the series $B(z) \equiv \sum_n c_n z^n/n!$. One can show that the "inverse" Borel transform that reconstructs $F(\lambda)$ from $B(z)$ is given by $\lambda F(\lambda) = \int_0^\infty \exp(-z/\lambda) B(z) dz$.
56. This tolerant view of non-renormalizable theories can be found in, e.g., Manohar (1997: 322) and Burgess (2007: 349).

2

The Role of Relativity in the CPT and Spin–Statistics Theorems

Naively, one might expect that the reason CPT invariance and the spin–statistics connection (SSC) are derivable properties in relativistic quantum field theories (RQFTs), but not in non-relativistic quantum field theories (NQFTs) and non-relativistic quantum mechanics (NQM), is due to relativity. In fact, an influential claim in the physics literature maintains that a violation of CPT invariance in an interacting RQFT entails a violation of Lorentz invariance (Greenberg, 2002). But Chapter 1 just got through telling us that relativity is neither necessary nor sufficient to derive either CPT invariance or the SSC in RQFTs, appropriately construed.

Of course, unpacking the role that relativity plays in the CPT and spin–statistics theorems requires specifying what we mean by "relativity." If we mean a strict sense of "invariance under restricted Lorentz transformations" (i.e., restricted Lorentz invariance, or RLI), then Chapter 1 tells us that that relativity is neither necessary nor sufficient for CPT invariance or the SSC: it is not sufficient since all the approaches to the CPT and spin–statistics theorems reviewed in Chapter 1 require more than just the assumption of RLI; and it is not necessary insofar as the algebraic approach reviewed in Section 1.2.4 replaces RLI with *modular covariance* (MC).

In addition to RLI or MC, all the approaches in Chapter 1 can be viewed as adopting a notion of locality that, in general, prohibits influences from traveling faster than the speed of light. Three distinct notions of locality appear in these approaches: *local commutativity* (LC) in the Wightman axiomatic approach,[1] *cluster decomposition* (CD) in Weinberg's approach, and *causality* in the Lagrangian and algebraic approaches. One might thus be motivated to encompass both RLI and "locality" under the rubric "relativity." But even this looser notion of relativity fails to explain why CPT invariance and the SSC are essential in RQFTs but not in NQFTs and NQM, for, again, as Tables 1.1 and 1.2 in Section 1.3 indicate, such a looser notion is neither necessary nor sufficient to derive these properties in RQFTs.

This chapter seeks to understand the role that relativity, in either the strict or the loose sense, plays in the CPT and spin–statistics theorems. Section 2.1

CPT Invariance and the Spin–Statistics Connection. Jonathan Bain.

considers the sense in which relativity is not sufficient for CPT invariance and the SSC by unraveling the relations between the different notions of locality, on the one hand, and RLI on the other. Section 2.2 considers the sense in which relativity is not necessary for CPT invariance and the SSC by unpacking the relations among RLI, MC, and variants of the latter.

The last part of Chapter 2 returns to Greenberg's (2002) claim, mentioned above, which maintains that a violation of CPT invariance in an *interacting* RQFT entails a violation of RLI. This claim has been influential in both the physics and philosophy of physics literature, since it suggests a test for violations of Lorentz invariance via experiments that measure CPT violation (Hagar, 2009: 261; Kostelecky, 2011). Moreover, in apparently linking Lorentz invariance with CPT invariance, it suggests the latter is somehow mysterious; thus, some philosophers have wondered how the charge conjugation transformation C can arise from a purely spatiotemporal symmetry (Arntzenius and Greaves, 2009; Greaves, 2010; Arntzenius, 2011). Greenberg's claim is puzzling, however, since, as Chapter 1 demonstrates, (restricted) Lorentz invariance is neither necessary nor sufficient for CPT invariance. Moreover, as argued in Chapter 1, standard proofs of the CPT theorem hold for *non-interacting* and, at most, *unrealistic interacting* states. Thus, on the surface, Greenberg's claim seems to both simplify the assumptions needed to derive CPT invariance, *and* address the issue of the extent of its applicability, all in one fell swoop. Section 2.3 offers a critique of Greenberg's claim in the context of the debate between what Chapter 1 refers to as pragmatism and purity.

2.1 Relativity and Locality

To set the stage for the discussion of the relations between RLI, on the one hand, and LC, *causality*, and CD, on the other, I would first like to consider the sense in which the latter constraints can be thought of as locality constraints. In the physics literature, they are typically motivated by an appeal to the requirement that spacelike separated events cannot influence each other, although the sense of "influence" differs among authors. For instance, Haag (1996) justifies the algebraic version of *causality* in the following way:

> ... the measurement of the field in a spacetime region $\mathcal{O}_1$ cannot perturb a measurement in a region $\mathcal{O}_2$ which is spacelike to $\mathcal{O}_1$ because there can be no causal connection between events in the two regions. Therefore such observables are compatible (simultaneously measureable). (Haag, 1996: 44)

Justifications for LC and *causality* follow a similar pattern:

> The justification for our postulate [viz., *causality*] lies in the fact that measurements at two space points with a spacelike distance can never disturb each other, since no signals can be transmitted with velocities greater than that of light. (Pauli, 1940: 721)

> The condition [LC] is often described as a *causality* condition, because if $x - y$ is spacelike then no signal can reach y from x, so that a measurement of ϕ at point x should not be able to interfere with a measurement of $\phi^\dagger$ at point y. (Weinberg, 1995: 198, slight notation change)

> ... we will define microscopic causality [viz., *causality*] as the statement that information cannot travel faster than the speed of light. For field theory, this means that $\phi(x)$ and $\phi(y)$ cannot interact with each other if they are separated by spacelike distances. Mathematically, this means that the commutator between these two fields must vanish for spacelike separations. (Kaku, 1993: 76)

> The commutativity of local fields at spacelike separated points... is a natural implementation at the quantum level of our macroscopic intuition that propagation of physical effects at superluminal speeds should be prohibited... (Duncan, 2012: 159)

> To really discuss causality, however, we should ask not whether particles can propagate over spacelike intervals, but whether a *measurement* performed at one point can affect a measurement at another point whose separation from the first is spacelike. The simplest thing we could try to measure is the field $\phi(x)$, so we should compute the commutator $[\phi(x), \phi(y)]$; if this commutator vanishes, one measurement cannot affect the other. (Peskin and Schroeder, 1993: 28)

Note that various notions of "influence" appear in these claims: Haag (1996), Pauli (1940), Weinberg (1995), and Peskin and Schroeder (1993) appeal to the independence of measurements of spacelike separated events. Haag (1996) links this independence with a prohibition on causal connections between such events, whereas Pauli (1940) and Weinberg (1995) link it with a prohibition on superluminal signaling, and Kaku (1993) links it with a prohibition on superluminal information transfer. Note finally, that whereas Duncan (2012) suggests that the independence of spacelike separated events be linked with a prohibition on superluminal propagation of physical effects, Peskin and Schroeder (1993) suggest the emphasis should instead be placed on independence of measurements.

There is a similar assortment of justifications for CD:

> It is one of the fundamental principles of physics (indeed, of all science) that experiments that are sufficiently separated in space have unrelated results. The probabilities for various collisions measured at Fermilab should not depend on what sort of experiments are being done at CERN at the same time. If this principle were not valid, then we could never make any predictions about any experiment without knowing everything about the universe. (Weinberg, 1995: 177)

> [CD is]... intimately linked to (but not synonymous with!) a special property of 'locality'... which will ensure, among other things, the 'Einstein causality' of the theory: namely, the absence of faster-than-light propagation of physical signals or effects. (Duncan, 2012: 58)

> The reason for this asymptotic factorization is that the correlation between two compatible observables A, B in a state is measured by the difference $\langle AB \rangle - \langle A \rangle \langle B \rangle$ where the bracket denotes the expectation value in the state. We expect that in the vacuum state the correlations of quantities relating to different regions decrease to zero as the separation of the regions increases to spacelike infinity. (Haag, 1996: 62)

> [The cluster property] ... is related to the independence of events associated to two clusters at infinite spacelike distance; the rate by which such a decorrelation is reached has actually been related to the decay rate of the interaction strength, or 'force' between the two clusters. (Strocchi, 2013: 73–74)

> The physical meaning of the cluster decomposition property is that, when two systems at points x and y become separated by a large spacelike distance, the interaction between them falls off to zero. (Streater and Wightman, 1964: 113)

In these quotes, Haag (1996), Weinberg (1995), and Strocchi (2013) appeal to an independence of (measurements of) correlations between observables or events, Duncan (2012) appeals to prohibitions on superluminal propagation and superluminal signaling, and Strocchi (2013) and Streater and Wightman (1964) appeal to an independence of interactions.[2]

Thus LC, *causality*, and CD all act to impose an independence of spacelike separated events. This independence is cashed out in a number of different ways: causal independence; independence of measurements, experiments, or interactions; prohibitions on superluminal signaling, superluminal propagation, or superluminal information transfer. This notion of locality as an independence condition should be contrasted with other notions that can be associated with RQFTs. For instance, within the algebraic approach, Earman and Valente (2014: 3) identify four senses of locality:[3]

(1) Localization: a requirement that the physical quantities of interest must be localized in finite regions of spacetime.
(2) Separability/independence conditions.
(3) No superluminal signaling (NSS) and/or no superluminal propagation (NSP).
(4) A notion of locality that is violated by correlations exhibited by subsystems of entangled states.

Note that this taxonomy is not specific to the algebraic approach; it is general enough to be applicable to notions of locality that appear in the approaches to the CPT and spin–statistics theorems reviewed in Section 1.2. Note, too, that in the context of the CPT and spin–statistics theorems, sense (4) can be ignored, since it plays no role in any of the proofs of these theorems.[4] Sense (1) appears as either an explicit or implicit assumption in these proofs, in addition to LC, *causality*, or CD. For instance, in the algebraic approach, Earman and Valente

(2014) consider sense (1) to be encoded in the correspondence between local algebras of observables and spacetime regions that defines a net of local algebras $\mathcal{O} \mapsto \mathfrak{R}(\mathcal{O})$. In the Lagrangian approach, sense (1) is represented by the constraint that requires the Lagrangian density to be a functional of fields and their derivatives evaluated at the same point, and a similar constraint appears implicitly in Weinberg's approach in the context of a theory's Hamiltonian. Similarly, the definition of Wightman fields in the axiomatic Wightman approach in terms of operator-valued distributions is a locality concern associated with sense (1).

On the other hand, senses (2) and (3) appear explicitly in the quotes that motivate LC, *causality*, and CD. Indeed, Earman and Valente identify sense (2) as the notion captured by *algebraic causality* (which they refer to as "microcausality"). Moreover, they argue that *algebraic causality* (and other similar separability/independence constraints) should be thought of as "setting the kinematics of the theory" (Earman and Valente, 2014: 3), i.e., those aspects of a theory that are independent of the specification of its dynamics. This is evident in the algebraic approach in which independence constraints like *algebraic causality* are imposed independently of the dynamics.[5] This is also evident in the axiomatic Wightman approach, the Lagrangian approach, and Weinberg's approach, in which LC, *causality*, and CD are likewise imposed independently of the dynamics. Earman and Valente then point out that, within the algebraic approach, it can be shown that *algebraic causality* is equivalent to NSS (under a particular notion of the latter in terms of non-selective measurements). Thus they view NSS as a purely kinematical feature of a theory and contrast it with NSP, which they view as a purely dynamical feature, depending in particular, on a theory's equations of motion.

These considerations suggest that LC, *causality*, and CD can be considered locality constraints in the sense of (2) above; namely, in the sense of being independence constraints that are kinematical, not dynamical, and that are motivated, at best, by NSS, as opposed to NSP.[6] As kinematical constraints, they are imposed on an RQFT prior to the articulation of the theory's dynamics. This makes sense if we understand CPT invariance and the SSC as kinematical properties, i.e., properties that a theory possesses regardless of the details of the dynamics it possesses. Moreover, as kinematical constraints, LC, *causality*, and CD say nothing about superluminal propagation, which, following Earman and Valente, is a property associated with the dynamics of a theory.

Now, to the extent that neither NSS nor NSP is entailed by RLI, we should not be surprised that LC, *causality*, and CD are independent of RLI.[7] In the following sections, this will be made more explicit.

2.1.1 Local Commutativity

Recall from Section 1.1.2 that *local commutativity* (LC) requires field operators to commute or anticommute:

$$[\phi(x), \phi^{\dagger}(y)]_{\mp} = 0, \text{ for spacelike } (x-y), \tag{LC}$$

where "$\mp$" indicates a commutator or anticommutator. A slightly stronger version of LC appears in the Wightman axiomatic proof of the spin–statistics theorem, namely, StLC (the *statistics–locality connection*), which requires field operators to commute or anticommute at spacelike separations, depending on whether they are bosonic or fermionic, respectively (Section 1.2.1). This is how the axiomatic approach encodes statistics on fields. Note that StLC entails LC. Moreover, LC, and not StLC, is one of the Wightman axioms, and RLI is another, separately posited, axiom. These axioms are independent in so far as one can construct non-local fields that satisfy RLI and violate LC (Streater and Wightman, 1964: 105).

More, however, can be said about the relation between LC and RLI. Note first that the basic objects in the Wightman axiomatic approach are Wightman functions defined as vacuum expectation values of unordered products of fields. Recall, however, that another type of object arose in Section 1.4.1 of Chapter 1 in the context of interacting fields; namely, vacuum expectation values of *time-ordered* products of fields, or "τ-functions." One might wonder if the relation between LC and RLI is any different for τ-functions. In fact, it is. One can demonstrate the following entailments:

(A) [(LC of fields) & (RLI of fields)] $\Rightarrow$ (RLI of time-ordered products of fields)

(B) [(RLI of τ-functions) & (RLI of Wightman functions)] $\Rightarrow$ (LC of fields)

Greenberg (2006a) notes that the first entailment is generally accepted and goes on to demonstrate the second entailment. He concludes with the following remarks:

> If we take Lorentz covariance of time-ordered products as the condition of Lorentz covariance of the field theory, then... local commutativity is not an independent assumption of the theory. This implies further that the spin–statistics and CPT theorems also hold without further assumptions. All of this means that relativistic quantum field theory becomes a more coherent structure. (Greenberg, 2006a: 1)

To assess this claim, we first have to assume it is referring to the Wightman axiomatic proofs of the spin–statistics and CPT theorems: LC does not occur in any of the other formulations reviewed in Chapter 1 (viz., Weinberg's approach, the Lagrangian approach, and the algebraic approach). Note, too, that StLC appears explicitly in the Wightman axiomatic proof of the Spin–Statistics theorem, whereas a weaker version (*weak local commutativity*, or WLC) appears in the axiomatic proof of the CPT theorem. On the other hand, LC entails WLC.

The first part of Greenberg's claim thus amounts to the following: in the axiomatic proofs of the spin–statistics and CPT theorems (Section 1.2.1), if we replace RLI of fields with RLI of τ-functions and Wightman functions, then we can drop the assumptions of LC and WLC.[8] In the remainder of this subsection, I will first review how entailments (A) and (B) are justified, and then consider the second part of Greenberg's claim; namely, whether replacing RLI of fields with RLI of τ-functions and Wightman functions leads to a "more coherent structure" for RQFTs.

Entailment (A) is based on the fact that the time-ordering of two spacetime points is RLI unless the points are spacelike separated.[9] Thus if a field $\phi(x)$ is RLI, then so are time-ordered products of $\phi(x)$, except when $\phi(x)$ is evaluated at spacelike separated points. But if $\phi(x)$ obeys LC, then time-ordering will not violate RLI even at such points.[10]

Greenberg's argument for entailment (B) is as follows. Recall from Section 1.2.1 that a Jost point $(x_1,\ldots,x_n)$ is a convex set of real points that are spacelike separated from each other, i.e., $(x_i - x_j)$ is spacelike for all $x_i, x_j \in (x_1,\ldots,x_n)$, $i \neq j$. If we assume that τ-functions and Wightman functions are RLI, then at a Jost point, they are equivalent (since time-ordering fails for Lorentz-invariant objects at Jost points). Moreover, one can choose a Lorentz transformation $\Lambda \in L_+^\uparrow$ that leaves the Jost point unchanged, except for reversing the time order of two of its points, call them x_ℓ and $x_{\ell+1}$. Thus $(\Lambda x_\ell)^0 > (\Lambda x_{\ell+1})^0$ and $\Lambda x_i = x_i$ otherwise, for all $x_i \in (x_1,\ldots,x_\ell,x_{\ell+1},\ldots,x_n)$, $i \neq \ell$ or $\ell+1$. Thus, at a Jost point,

$$
\begin{aligned}
W^{(n)}(x_1,\ldots,x_\ell,x_{\ell+1},\ldots,x_n) &= \langle 0|\phi(x_1)\ldots\phi(x_\ell)\phi(x_{\ell+1})\ldots\phi(x_n)|0\rangle \\
&= \langle 0|\,T\{\phi(x_1)\ldots\phi(x_\ell)\phi(x_{\ell+1})\ldots\phi(x_n)\}|0\rangle \\
&= \langle 0|T\{\phi(\Lambda x_1)\ldots\phi(\Lambda x_\ell)\phi(\Lambda x_{\ell+1})\ldots\phi(\Lambda x_n)\}|0\rangle \\
&= \pm\langle 0|\phi(x_1)\ldots\phi(x_{\ell+1})\phi(x_\ell)\ldots\phi(x_n)|0\rangle
\end{aligned}
$$

The first line is the definition of a Wightman function, the second line follows from the equivalence of Wightman functions and τ-functions at Jost points, the third line follows from RLI of τ-functions, and the fourth line follows from the definition of Λ (where the sign "$\pm$" depends on whether the fields commute or anticommute).[11] It now follows that $\langle 0|\phi(x_1)\ldots[\phi(x_\ell),\phi(x_{\ell+1})]_\mp\ldots\phi(x_n)|0\rangle = 0$. Greenberg then appeals to Wightman's (1956) reconstruction theorem to justify LC (which, in this case, is the condition $[\phi(x_\ell),\phi(x_{\ell+1})]_\mp = 0$, for any spacelike separated points x_ℓ and $x_{\ell+1}$).

Does replacing RLI for fields with RLI for τ-functions and Wightman functions lead to a more coherent structure for RQFTs, as Greenberg suggests? Note first that while there is a Reconstruction theorem for Wightman functions (Wightman, 1956) that, ultimately, justifies replacing RLI of fields with RLI of Wightman functions, no similar reconstruction theorem holds for τ-functions, as Greenberg (2006a: 2, footnote 2) points out.[12] On the other hand, recall from Chapter 1 that axiomatic approaches to RQFTs face the *existence problem*, namely, there are no models of the axioms of the reconstructed field theory for *realistic* interacting theories. Note, furthermore, that typical pragmatist approaches employ τ-functions in their perturbative approaches to realistic interacting theories (Section 1.4.1). This suggests that adopting τ-functions may lead, as Greenberg suggests, to a "more coherent structure" for *interacting* RQFTs.[13] This will be considered in greater detail in Section 2.3.

2.1.2 Causality

Recall from Section 1.2.3 that *causality* requires:

The observable quantities associated with an RQFT commute at spacelike distances (Causality)

In the algebraic approach (Section 1.2.4), this becomes:

For $A_1 \in \mathfrak{R}(\mathcal{O}_1)$, $A_2 \in \mathfrak{R}(\mathcal{O}_2)$, *and* $\mathcal{O}_1$, $\mathcal{O}_2$ *spacelike separated*, $[A_1, A_2] = 0$. (Algebraic causality)

for a net of von Neumann algebras $\mathcal{O} \mapsto \mathfrak{R}(\mathcal{O})$. (It should be noted that some authors refer to both of these constraints as "micro-causality.")

In the Lagrangian approach, RLI does not entail *causality* insofar as one can construct Lagrangians that satisfy RLI but violate *causality*. Conversely, *causality* does not entail RLI insofar as one can construct non-relativistic Lagrangians from fields that satisfy *causality* (appropriately construed in the non-relativistic context). On the other hand, entailment (A) of Section 2.1.1 indicates that *causality* and RLI of observables is sufficient for time-ordered products of observables to be RLI.

In the algebraic approach, RLI does not entail *algebraic causality*. One way to demonstrate this would assumedly be to recast examples of Lagrangians that satisfy RLI but violate *causality* as models of the algebraic axioms. Another way is suggested by Earman and Valente (2014). They review how *algebraic causality* is equivalent to a prohibition on superluminal signaling (i.e., NSS) by representing a signaling process in terms of a non-selective measurement. On the strength of this equivalence, since RLI does not entail NSS, RLI does not entail *algebraic causality*.

Causality and Local Commutativity

In an insightful article, Greenberg (1998) suggests that LC and *causality* are ontologically distinct assumptions, the former associated with fields, and the latter associated with particles. In particular, Greenberg distinguishes between two theorems: the "spin–statistics" theorem, which states that "... *particles* that obey Bose statistics must have integer spin and *particles* that obey Fermi statistics must have odd half-integer spin," and the "spin–locality" theorem, which states that "... *fields* that commute at spacelike separation must have integer spin and *fields* that anticommute at spacelike separation must have odd half-integer spin" (Greenberg, 1998: 144). Greenberg identifies the spin–statistics theorem with the proofs due to Fierz (1939) and Pauli (1940) that inform the textbook Lagrangian approach,[14] and takes the *causality* assumption to be the requirement that

"...the bilinear observables constructed from the (free) asymptotic fields [i.e., the in- and out-fields of Haag–Ruelle scattering theory] commute at spacelike separation" (Greenberg, 1998: 145). He identifies the spin–locality theorem with the proofs due to Lüders and Zumino (1958) and Burgoyne (1958) that inform the axiomatic approach, and distinguishes it from the spin–statistics theorem solely on its replacement of *causality* with LC.[15] Moreover, he claims that these theorems differ on what they say "fails" for the "wrong" cases:

> Spacelike commutativity (locality) of observables fails for the wrong cases of the spin–statistics theorem. (Greenberg, 1998: 144)

whereas

> The Lüders–Zumino and Burgoyne proof shows that if the fields have the wrong commutation relations, i.e., integer-spin fields are antilocal and odd-half-integer-spin fields are local, the fields vanish. This assumption does not relate directly to particle statistics and for that reason this theorem should not be called the spin–statistics theorem. (Greenberg, 1998: 146)

In these quotes, I take "spacelike commutativity (locality) of observables" to be the assumption of *causality*, and the condition that "the fields have the wrong commutation relations" I take to be NSpLC, i.e., the *wrong spin–locality connection* of Section 1.1.2. Thus to reiterate, Greenberg makes two claims:

(i) The axiomatic theorem should be called the spin–locality theorem, whereas the textbook Lagrangian theorem should be called the spin–statistics theorem.
(ii) The axiomatic theorem differs from the textbook Lagrangian theorem only in replacing *causality* with LC.

To further unpack these claims, consider the reconstructions of the axiomatic (A1) and Lagrangian (C1, C2) proofs of the spin–statistics theorem in Sections 1.2.1 and 1.2.3, reproduced below:

A1. [(RLI for fields) & SC & StLC] $\Rightarrow$ (SSC for non-vanishing fields)
C1. [(RLI of fields) & SC & (1.1) & *causality*] $\Rightarrow$ [SSC for fermionic fields]
C2. [(RLI of fields) & (1.1) & *causality*] $\Rightarrow$ [SSC for bosonic fields]

For ease of comparison, we might collapse C1 and C2 onto C:

C. [(RLI of fields) & SC & (1.1) & *causality*] $\Rightarrow$ [SSC for fields]

since C2 certainly entails that a positive-energy relativistic field that satisfies condition (1.1) of Section 1.1.2 possesses the SSC. But A1 cannot be obtained from

C by replacing *causality* with LC. In fact, LC does not even occur in A1! Moreover, neither A1 nor C refers explicitly to the spin–locality connection (SpLC). Obviously A1 and C are not the theorems Greenberg has in mind.

Note, however, that in the axiomatic approach, one can show that the following holds:[16]

A1′. [RLI & SC & LC] $\Rightarrow$ (~NSpLC $\vee$ fields vanish)

In the Lagrangian approach, one can derive a similar entailment:[17]

C′. [RLI & SC & *causality*] $\Rightarrow$ ~(NSpLC for a, $a^{\dagger}$)

Entailment A1′ says that a relativistic positive-energy field that satisfies LC either fails to satisfy the wrong spin–locality connection or vanishes. Entailment C′ says that the creation and annihilation operators of a relativistic positive-energy field that satisfies *causality* fail to satisfy the wrong spin–locality connection. These entailments seem close to Greenberg's claim (ii); namely, they differ in their assumptions only in that whereas entailment A1′ appeals to LC, entailment C′ appeals to *causality*. Moreover, these claims also differ on what they say "fails" for the "wrong" cases: suppose the "wrong" case is NSpLC (i.e., the *wrong spin–locality connection*). Then C′ implies that if the "wrong" case is assumed, then (if we maintain RLI and SC) *causality* must fail, whereas A1′ implies that if the "wrong" case is assumed, then (if we maintain RLI and SC and LC), the fields must vanish. Note finally that A1′ as it stands says nothing about spin and statistics, and thus, as Greenberg suggests, might more appropriately be called the spin–locality theorem. It is a claim about the conditions under which the *spin–locality connection* (SpLC) holds, as opposed to the SSC. But C′ also says nothing about spin and statistics, and should rather be viewed similarly as a claim referring to SpLC and not SSC. It might thus be more appropriate to designate both A1′ and C′ as axiomatic and Lagrangian versions of the spin–locality theorem. This would allow one to maintain the view that A1 and C are axiomatic and Lagrangian versions of the spin–statistics theorem.

Thus while Greenberg's distinction between the spin–statistics theorem and the spin–locality theorem, based on the distinction between SSC and SpLC, is deeply insightful, his claims (i) and (ii) are slightly misleading. Claim (i) fails to recognize that there is an axiomatic version of the spin–statistics theorem, and there is a Lagrangian version of the spin–locality theorem.[18] Claim (ii) fails to recognize that it is only for the spin–locality theorem that the axiomatic version differs from the Lagrangian version in the replacement of *causality* with LC.

On the other hand, Greenberg might argue that SpLC for a, $a^{\dagger}$ is synonymous with SSC, in which case the consequent of C′ would be ~NSSC, and thus C′ could be interpreted as the spin–statistics theorem, with A1′ being the spin–locality theorem. The argument might be that statistics is implicitly encoded in the (anti-)commutation relations of creation and annihilation operators, thus attributing spin to the latter (i.e., asserting SpLC for a, $a^{\dagger}$) amounts to an assertion of SSC. In fact, this is one way to interpret Greenberg's claim that statistics has to do with particles and not fields. This is a viable position, but it requires, among other things, an implicit identification of particle states with Fock space states, which leads to problems in maintaining a particle interpretation in realistic interacting RQFTs (see, e.g., Fraser, 2008). More importantly, however, I think it fails to recognize StLC as an assumption in the axiomatic spin–statistics theorem.

At this point it may be helpful to compare the various connections between spin, statistics, and commutation relations that have appeared in this and the previous chapter. These are reproduced in Table 2.1. SSC is the *spin–statistics connection*, which says integer-spin states possess Bose–Einstein (BE) statistics, and half-integer-spin states possess Fermi–Dirac (FD) statistics. The *wrong spin–statistics connection* is represented by NSSC. SpLC is the *spin–locality connection*, which says integer-spin states commute at spacelike separations, and half-integer-spin states anticommute at spacelike separations. The *wrong spin–locality connection* is represented by NSpLC. Finally, StLC is the *statistics–locality connection*, which says states that possess BE statistics commute at spacelike separations, and states that possess FD statistics anticommute at spacelike separations. The *wrong statistics–locality connection* is represented by NStLC.

Note that, in general, a failure of a connection (i.e., "~[*connection*]") is not the same as the wrong connection, insofar as a failure of a state to commute (resp. anticommute) is not synonymous with the state anticommuting (resp. commuting). In the case of SSC and NSSC, however, one can argue that ~NSS is equivalent to SSC, provided we allow that any theory with parastatistics is a notational variant of a theory with normal (BE or FD) statistics (Baker et al., 2014).

With the distinctions in Table 2.1, we can describe the following relations between *causality* and LC, SpLC, and StLC. As noted in Section 1.2.3, how one interprets *causality* will depend on what one understands the observables of a theory to be. Thus if the observables of a theory are the integer-spin fields or bilinears in half-integer-spin fields that appear in the theory, then the SpLC entails

Table 2.1 *Connections between spin, statistics, and (anti-)commutation relations.*

Connection	Wrong connection
SSC: Integer-spin states possess BE statistics; half-integer-spin states possess FD statistics.	NSSC: Integer-spin states possess FD statistics; half-integer-spin states possess BE statistics.
SpLC: Integer-spin states commute; half-integer-spin states anticommute.	NSpLC: Integer-spin states anticommute; half-integer-spin states commute.
StLC: States that possess BE statistics commute; states that possess FD statistics anticommute.	NStLC: States that possess BE statistics anticommute; states that possess FD statistics commute.

causality. This is the implicit understanding of observables in the Lagrangian version C of the spin–statistics theorem (as noted in Section 1.2.3), and in the Lagrangian version C′ of the spin–locality theorem. On the other hand, if the observables of a theory are taken to be the bosonic fields or bilinears in fermionic fields that appear in the theory, then the StLC entails *causality*. Finally, if the observables of a theory are taken to be the states that commute, or bilinears in states that anticommute, then LC entails *causality*.

Table 2.1 also suggests various relations between LC, SpLC, and StLC on the one hand, and the condition of WLC in the axiomatic proof of the CPT theorem (Section 1.2.1). WLC requires that, in the neighborhood of a Jost point,

$$\langle 0 | \phi(x_1) \dots \phi(x_n) | 0 \rangle = i^F \langle 0 | \phi(x_n) \dots \phi(x_1) | 0 \rangle \qquad \text{(WLC)}$$

where F, in general, is the number of anticommuting fields in the product of fields. In this general case, WLC is a weaker version of LC, in the sense that the latter entails the former. On the other hand, some authors (Streater and Wightman, 1964: 131, 145) identify F as the number of half-integer-spin fields, which implies that WLC is a weaker version of SpLC. Other authors (Haag, 1996: 98; Greenberg, 2006b: 1547), identify F as the number of fermionic fields, which implies WLC is a weaker version of StLC. But note that in the last case, WLC cannot be thought of as a weaker version of SpLC, since it says nothing about spin. Thus Greenberg's (2006b: 1546) assertion that WLC "... is implied by the spin–locality theorem but is weaker than that theorem since we need this relation [viz., WLC] only at Jost points..." appears to be inconsistent with his identification of F with the number of fermionic fields.

2.1.3 Cluster Decomposition

Cluster decomposition (CD) of the S-matrix is one of the assumptions in Weinberg's formulation of the spin–statistics and CPT theorems. Recall from Section 1.2.2 that it requires the S-matrix to encode the requirement that scattering experiments in regions of spacetime that are separated by large spatial distances be independent of each other:

> *Let $S_{\beta_1+\ldots+\beta_N,\alpha_1+\ldots+\alpha_N}$ represent the S-matrix for N multi-particle processes $|\alpha_1\rangle \to |\beta_1\rangle, \dots, |\alpha_N\rangle \to |\beta_N\rangle$. If all particles in states $|\alpha_i\rangle, |\beta_i\rangle$ are at a great spatial distance from all particles in states $|\alpha_j\rangle, |\beta_j\rangle$, for $i \neq j$, then the S-matrix factorizes: $S_{\beta_1+\ldots+\beta_N,\alpha_1+\ldots+\alpha_N} = S_{\beta_1\alpha_1} \dots S_{\beta_N\alpha_N}$.* (CD)

The significance of CD in Weinberg's approach is that it is a necessary condition for the Hamiltonian, which encodes the dynamics of a theory, to take a particular general form (in fact Weinberg, 1995: 182 suggests that it is sufficient, too). On first blush, this may seem to contradict what was said above in the introduction to Section 2.1 about the nature of locality constraints like CD. It was argued there that they are *kinematical* constraints on a theory, as opposed to *dynamical* constraints. In the following, I will first review the argument that leads from CD to a theory's Hamiltonian, and then assess the relation between CD and RLI. I will then consider the extent to which CD amounts to a kinematical, as opposed to a dynamical, constraint.

Duncan (2012: 132) describes two intuitions about scattering interactions that underwrite CD. The first involves a spatial constraint that requires that the total scattering amplitude for two groups of particles associated with distinct finite spatial regions of spacetime should approach the product of the independent scattering amplitudes as the distance between the regions goes to infinity. This encodes the intuition that scattering events at great spatial distances should be independent of each other. The second intuition describes a momentum conservation constraint that requires that, if no interaction occurs between two groups of particles, the total momentum of each group should be preserved; whereas if interactions do occur, the probability for individual momenta to be exactly conserved should be zero. Informally, as Duncan suggests, these intuitions are related by a Fourier transformation: the spatial clustering constraint of the first intuition transforms, under the second intuition, into a momentum conservation constraint. The latter can be represented mathematically by requiring that the terms in the expansion of the S-matrix that describe total interaction between particle subgroups (i.e., the "connected" terms) be associated with a momentum conserving Dirac delta function. Moreover, as it turns out, this momentum conservation constraint on the connected terms of the S-matrix can be shown to be generated by a similar momentum conservation constraint imposed on the theory's Hamiltonian. This is formalized by the following claim (Weinberg, 1995: 182; Duncan, 2012: 149):

Theorem 2.1 Let multi-particle state β be comprised of N particles with ith momentum $\mathbf{p}'_i$ and multi-particle state α be comprised of M particles with ith momentum $\mathbf{p}_i$. Then the S-matrix $S_{\beta\alpha}$ satisfies CD if the Hamiltonian H takes the following form:

$$H = \sum_{N,M=0}^{\infty} \int dq'_1 \ldots dq'_M dq_1 \ldots dq_M a^\dagger(q'_1) \ldots \\ a^\dagger(q'_M) a(q_M) \ldots a(q_1) h_{NM}(q'_1 \ldots q_M) \tag{2.1}$$

where the coefficient functions h_{NM} contain a single momentum conservation δ-function:

$$h_{NM}(q'_1, \ldots, q'_N, q_1, \ldots, q_M) = \delta^3(\mathbf{p}'_1 + \ldots + \mathbf{p}'_N - \mathbf{p}_1 - \ldots - \mathbf{p}_M) \\ f_{NM}(q'_1, \ldots, q'_N, q_1, \ldots, q_M)$$

where f_{NM} are smooth functions of the momenta (in the sense of not depending on additional delta functions of momenta).[19]

Theorem 2.1 says that a sufficient condition for CD of the S-matrix of a theory is that the theory's Hamiltonian take the form of a sum of creation and annihilation operators with coefficient functions h_{NM} that contain just a single three-dimensional (3-dim) momentum conservation delta function. The proof involves the following steps (as reconstructed in Bain, 1999: 21):

1. One first shows that CD is equivalent to the condition that the "connected" matrix components of the S-matrix, call them $S^C_{\beta\alpha}$, vanish as the spatial separation between clusters goes to infinity. The connected parts of the S-matrix are terms that encode a total interaction between all particles concerned. In general, in a scattering process involving some number N of particles, all of them can interact, or various subgroups may interact while others do not. The total scattering amplitude is the sum of the amplitudes that describe all the possible ways the particles may or may not interact. The connected parts of this total sum are the terms that encode the cases of total interaction between subgroups. If a connected term goes to zero, this means the groups of particles it represents do not interact.
2. One then demonstrates that $S^C_{\beta\alpha} \to 0$ as the spatial separation between clusters goes to infinity *if and only if* $S^C_{\beta\alpha}$ contains a *single* momentum-conservation δ-function factor $\delta^3(\mathbf{p}_\beta - \mathbf{p}_\alpha)$, where $\mathbf{p}_\beta(\mathbf{p}_\alpha)$ is the sum of all particle momenta in the state $\beta(\alpha)$. (A proof is given in Bain, 1999: 24.)
3. One then demonstrates that if the connected matrix components $H^C_{\beta\alpha}$ of the Hamiltonian contain a single factor of $\delta^3(\mathbf{p}_\beta - \mathbf{p}_\alpha)$, then so does $S^C_{\beta\alpha}$ (Weinberg, 1995: 186–187; Duncan, 2012: 140–143). Key to this proof is the identification of the connected components of the Hamiltonian H (which describe particle interactions in which all particles participate) with the matrix components of the interaction part H_{int} of H.
4. One then demonstrates that $H^C_{\beta\alpha} = \langle q'_1 \ldots q'_N | H^C | q_1 \ldots q_M \rangle = h_{NM}$.[20]

By assumption, the expansion coefficient functions h_{NM} of the Hamiltonian contain a single delta momentum conservation delta function. By (4), so do the connected matrix components $H^C_{\beta\alpha}$. By (3), so do the connected matrix components $S^C_{\beta\alpha}$ of the S-matrix. By (2), this means $S^C_{\beta\alpha} \to 0$ at great spatial separation between clusters; and by (1), this entails CD.

CD and RLI

CD of the S-matrix does not entail it satisfies RLI insofar as one can construct non-relativistic S-matrices that satisfy CD. Moreover, RLI does not entail CD: one can construct S-matrices that satisfy RLI but violate CD. The following example is given by Duncan (2012: 123–124) and Weinberg (1995: 187–188). Let the interaction Hamiltonian be given by:

$$H_{int} = \sum_{N=2}^{\infty} \int \prod_{i=1}^{N} \frac{d^3 q_i}{2E(\mathbf{p}_i)} \frac{d^3 q'_i}{2E(\mathbf{p}'_i)} \delta^4\left(\sum_i q'_i - \sum_i q_i\right) h^{(N)}(q'_i, q_i) |q'_1 \ldots q'_N\rangle\langle q_1 \ldots q_N| \tag{2.2}$$

This describes the scattering of an incoming N-particle state with momenta $q_1, \ldots, q_N$ into an outgoing N-particle state with momenta $q'_1, \ldots, q'_N$. If we require $h^{(N)} = \left[h^{(N)}\right]^*$ and that $h^{(N)}$ be a function of Lorentz invariant scalar products of the momenta, then H_{int} is unitary and satisfies RLI. But the corresponding S-matrix (given by equation 1.2 of Section 1.2.2) does not necessarily satisfy CD; in particular, we are free to choose the amplitudes $h^{(N)}$ such that they contain more than a single delta function. To see this, one can show that the amplitude for 3–3 scattering is given by

$$\langle q'_1 q'_2 q'_3 \,|\, H_{int} \,|\, q_1 q_2 q_3 \rangle = h^{(3)} + h^{(2)} \delta^3(\mathbf{p}'_3 - \mathbf{p}_3) \pm \text{permutations.} \tag{2.3}$$

Now suppose we wish to describe a scenario in which there is no scattering between three or more particles. This requires, among other things, that the above expression vanish; thus,

$$h^{(3)} = -h^{(2)} \delta^3(\mathbf{p}'_3 - \mathbf{p}_3) \mp \text{permutations.} \tag{2.4}$$

This violates CD in so far as the corresponding Hamiltonian now contains terms with coefficient functions that contain more than one delta function, since $h^{(2)}$ already has a factor of $\delta^3(\mathbf{p}'_1 + \mathbf{p}'_2 - \mathbf{p}_1 - \mathbf{p}_2)$. From the spatial intuition point of view, in this scenario, "... the introduction of a third particle localized arbitrarily far from two other localized interacting particles would have instantly caused these two to cease interacting..." (Duncan, 2012: 124).

There is an axiomatic version of CD that makes its relation to RLI a bit more explicit. Note, of course, that this *axiomatic CD* property is not directly relevant to proofs of the spin–statistics and CPT theorems, in so far as it does not appear in the axiomatic approach to these proofs. On the other hand, it may add some clarity to the relation between Weinberg's approach and the Wightman axiomatic approach. The axiomatic version of CD states the following (Streater and Wightman, 1964: 111):

> *If a is a spacelike vector, then*
> $W^{(n)}(x_1, \ldots, x_j, x_{j+1} + \lambda a, x_{j+2} + \lambda a, \ldots, x_n + \lambda a)$
> $\rightarrow W^{(j)}(x_1, \ldots, x_j) W^{(n-j)}(x_{j+1}, \ldots, x_n),$ (Axiomatic CD)
> *as* $\lambda \rightarrow \infty$ (in the sense of distributions).

Axiomatic CD plays a role in Wightman's (1956) reconstruction theorem which demonstrates how an RQFT satisfying the Wightman axioms can be reconstructed from its Wightman functions. In particular, *axiomatic CD* ensures that the reconstructed field theory possesses a unique vacuum state (Streater and Wightman, 1964: 110, 124). *Axiomatic CD* also plays a central role in the axiomatic approach to scattering given by Haag–Ruelle collision theory, in which it is required to demonstrate the existence of free asymptotic fields, and thus the existence of "a *non-perturbative* definition of the S-matrix" (Strocchi, 2013: 78;

see also page 123). Araki et al. (1962) show that *axiomatic CD* is entailed by the conjunction of the uniqueness of a translation invariant vacuum state, the *spectrum condition* (SC), and *causality*. In particular, RLI need not be assumed.

A Kinematical or a Dynamical Constraint?

I'd now like to consider whether the constraint imposed by CD on the form of the Hamiltonian in Weinberg's approach amounts to a *kinematical* or a *dynamical* constraint. The fact that the Hamiltonian of a theory encodes the theory's dynamics might suggest that CD imposes a dynamical constraint on a theory by restricting its Hamiltonian to have a specific form; namely, that given by equation (2.1). Duncan (2012: 137), for instance, views equation (2.1) as a constraint "...imposed on the dynamics of the theory (specifically, on the form of the Hamiltonian) by clustering..." In the introduction to Section 2.1, I claimed that CD is best understood as a kinematical constraint, i.e., a constraint imposed independently of the specification of a theory's dynamics. How can CD be thought of in this way if it constrains how a theory's dynamics, in the form of its Hamiltonian, is specified? The answer lies in the sense in which kinematical constraints are independent of dynamics. Note that the nature of the constraint on the Hamiltonian imposed by CD is formal rather than specific: it requires the Hamiltonian to take the general form of a sum of creation and annihilation operators, with coefficients that contain at most a single 3-dim momentum conservation delta function. Within this prescribed general form, there are still a large number of degrees of freedom, encoded in the arbitrariness of the functions f_{NM} in equation (2.1). Thus the general form of equation (2.1) remains unaltered under changes in the specific type of dynamics, as encoded in the f_{NM}. In other words, the constraint imposed by CD on a theory is robust under specific changes of the theory's dynamics. In this sense, CD is independent of the details of the dynamics and hence can be considered a kinematical constraint.

CD versus Local Commutativity and Causality

Finally, a few comments should be made about the relations among CD, LC, and *causality* in Weinberg's approach. Recall from Section 1.2.2 that for Weinberg, LC is a less fundamental constraint than CD. In particular, LC of fields is one of the conditions under which an S-matrix can be constructed that satisfies both RLI and CD. Likewise, *causality* is less fundamental than CD (see, e.g. footnote 14). *Causality*, in a rather restricted form, is one way to satisfy the constraint of RLI of the S-matrix.

2.2 Relativity and Geometric Modular Action

The previous section makes it explicit that relativity is not sufficient for CPT invariance and the SSC. This section investigates the sense in which relativity is not necessary to derive these properties, appropriately construed. The setting here is the algebraic approach to the spin–statistics and CPT theorems. Recall from Section 1.1 that the significance of RLI in derivations of SSC is that it is a way of encoding spin; namely, the spin of a state can be encoded in the representation of the covering of the restricted Lorentz group that leaves the state invariant. The significance of RLI in derivations of CPT invariance is that PT transformations are elements of the proper Lorentz group (thus if a state is invariant under the proper Lorentz group, it is invariant under PT transformations). In the algebraic approach, RLI is typically cashed out in terms of "Poincaré covariance," which is the requirement that the von Neumann algebra of observables $\mathfrak{R}$ admits a representation of the Poincaré group (which adds translations to the Lorentz group). In particular, to recover SSC, one would like to construct a representation of the covering of the restricted Poincaré group, and to recover CPT invariance, one would like to construct a representation of the proper Poincaré group. Recall from Section 1.2.4 that Guido and Longo's (1995) algebraic versions of the spin–statistics and CPT theorems demonstrate how representations of the Poincaré group can be constructed by replacing RLI with the assumption of MC. A slightly different algebraic approach replaces RLI with what Kuckert (1995) refers to as *modular P_1CT-symmetry* (MPCT). MPCT and MC are specific versions of two general constraints that Buchholz et al. (2000) refer to as the *condition of geometric modular action* (CGMA) and the *condition of geometric action for the modular groups* (CMG). This section investigates the relations among these algebraic conditions, and their relation to RLI.

2.2.1 Poincaré Covariance from Modular Data

Modular covariance (MC) and *modular P_1CT-symmetry* (MPCT) are two relations uncovered by Bisognano and Wichmann (1975) between algebraic structures associated with von Neumann algebras generated by Wightman fields on the one hand, and geometric properties of wedge regions of Minkowski spacetime on the other. These relations in turn are consequences of a more general analysis of the structure associated with von Neumann algebras initiated by Tomita and Takesaki. The Tomita–Takesaki theorem demonstrates that a von Neumann algebra $\mathfrak{R}$ of bounded linear operators on a Hilbert space $\mathcal{H}$ with a cyclic and separating vacuum vector Ω possesses a one-parameter collection of unitary operators $\Delta(t)$ (the "modular operators"), $t \in \mathbb{R}$, and an antiunitary operator $\mathcal{J}$ (the "modular conjugation") such that $\mathcal{J}\Omega = \Omega = \Delta^{it}\Omega$, $\Delta^{it}\mathfrak{R}\Delta^{-it} = \mathfrak{R}$, and $\mathcal{J}\mathfrak{R}\mathcal{J} = \mathfrak{R}'$, where $\mathfrak{R}'$ is the commutant of $\mathfrak{R}$ (i.e., the set of bounded linear operators that commute with all elements of $\mathfrak{R}$). Thus associated with the algebraic objects $(\mathfrak{R}, \Omega)$ are the "modular data" $(\Delta(t), \mathcal{J})$.

The Tomita–Takesaki theorem is a purely algebraic result. Bisognano and Wichmann (1975) gave it a geometric interpretation in the following way. They considered a net of von Neumann algebras $\mathcal{O} \mapsto \mathfrak{R}(\mathcal{O})$ generated by (finite component) Wightman fields that satisfy RLI, the SC, and LC. These conditions entail the cyclic vacuum state is separating (this is the content of the *separating corollary* discussed in the Appendix), and this allows the Tomita–Takesaki theorem to apply. They then considered the restrictions $\mathfrak{R}(W)$ of the local algebras to wedge regions W of Minkowski spacetime M, where a wedge region is any Poincaré transformation of the set $\{x \in M : x_1 > |x_0|\}$, where (x_0, x_1, x_2, x_3) is an inertial coordinate system. Bisognano and Wichmann showed that the modular operators Δ_W^{it} and modular conjugation $\mathcal{J}_W$ associated with $\mathfrak{R}(W)$ act on elements of $\mathfrak{R}(\mathcal{O})$ in the following ways:

$$\text{(a)}\quad \Delta_W^{it}\mathfrak{R}(\mathcal{O})\Delta_W^{-it} = \mathfrak{R}(\Lambda_W(t)\mathcal{O}) \qquad \text{(MC)}$$

$$\text{(b)}\quad \mathcal{J}_W\mathfrak{R}(\mathcal{O})\mathcal{J}_W = \mathfrak{R}(j\mathcal{O}) \qquad \text{(MPCT)}$$

where $\Lambda_W(t)$ is the one-parameter group of (1 + 1)-dim Lorentz boosts in the x_1 direction that leaves W invariant,[21] and j is a (1 + 1)-dim "P_1T" transformation on W, i.e., $j(x_0, x_1, x_2, x_3) = (-x_0, -x_1, x_2, x_3)$. Guido and Longo (1995) refer to condition (a) as MC, whereas Kuckert (1995) refers to condition (b) as MPCT. Bisognano and Wichmann also showed that (for scalar fields), the modular conjugation $\mathcal{J}_W$ is given by,[22]

$$\text{(c)}\quad \mathcal{J}_W = \Theta R_W$$

where Θ is the CPT operator associated with the Wightman fields, and R_W implements a rotation of π about the x_1-axis (R_W implements the "partial" parity transformation, call it $\mathsf{P}_{2,3}$, given by $\mathsf{P}_{2,3}(x_0, x_1, x_2, x_3) = (x_0, x_1, -x_2, -x_3)$).

Thus Bisognano and Wichmann demonstrated that, under the assumption that the local algebras $\mathfrak{R}(\mathcal{O})$ satisfy RLI, SC, and meet the requirements for the Tomita–Takesaki theorem, the modular operators and modular conjugations associated with wedge algebras implement (1 + 1)-dim Lorentz boosts on wedge regions, and (1 + 1)-dim spacetime reflections on wedge regions, respectively.

Poincaré Covariance from MC and MPCT

Guido and Longo's (1995) strategy in deriving the SSC and CPT invariance was to invert the reasoning in Bisognano and Wichmann. Guido and Longo assumed a net of local von Neumann algebras (not necessarily generated by Wightman fields) that meets the requirements of the Tomita–Takesaki theorem, and in addition satisfies MC, and nothing more; in particular, one does not assume RLI. They were then able to construct a unique positive energy unitary representation $g \in P_+^\uparrow \mapsto U_0(g)$ of the restricted Poincaré group $P_+^\uparrow$, under which (1 + 1)-dim Lorentz boosts on wedge regions are mapped to the corresponding modular operators: $U_0(\Lambda_W(t)) = \Delta_W^{it}$ (Guido and Longo, 1995: 530). CPT invariance follows by defining the CPT operator by inverting relation (c). To fully justify this, one needs to construct a representation of the proper Poincaré group P_+, and this is accomplished by extending U_0 by defining $U_0(I_W) = \mathcal{J}_W$, where I_W

is a P_1T transformation on the wedge W. The representation U_0 then induces a representation U_ρ of the covering $\tilde{P}_+^\uparrow$ of the restricted Poincaré group that acts on DHR representations ρ. One can then show that the SSC holds for U_ρ, i.e., for any DHR representation ρ, $U_\rho(2\pi) = \kappa_\rho$, where κ_ρ is the statistical phase of ρ (see Section 1.2.4).

In effect, Guido and Longo construct a unique positive energy representation of the (3 + 1)-dim Poincaré group that acts on a von Neumann algebra $\mathfrak{R}$, and that is necessary in deriving CPT invariance and the SSC, ultimately from a representation that implements (1 + 1)-dim Lorentz boosts by means of modular operators. A similar approach to deriving CPT invariance and the SSC from modular data, but this time using modular conjugations, was initiated by Kuckert (1995), extended in Kuckert (2005) and Kuckert and Lorenzen (2007), and reached fruition in Lorenzen (2007). The goal of these works is to construct a unique positive energy unitary representation of the proper Poincaré group (to recover a CPT operator), and a positive energy unitary representation of the covering of the restricted Poincaré group (to recover the SSC). However, instead of assuming MC, this approach assumes MPCT. Kuckert (1995) initially showed that a field algebra satisfying RLI, *normal commutation relations*, the SC, and MPCT possesses a CPT operator and the SSC. Kuckert (2005) showed how the assumptions of RLI and SC can be dropped by using MPCT to construct a representation of the covering of the rotation group $SO(3)$, and thus obtain CPT invariance and the SSC for a rotationally invariant QFT. Kuckert and Lorenzen (2007) extended this method to recover a representation of the covering $\tilde{L}_+^\uparrow$ of the restricted Lorentz group and Lorenzen (2007) recovered a representation of the covering $\tilde{P}_+^\uparrow$ of the restricted Poincaré group.[23] Thus whereas Guido and Longo obtain their (3 + 1)-dim representation of the Poincaré group from an implementation of (1 + 1)-dim Lorentz boosts, Kuckert et al. obtain theirs from an implementation of (1 + 1)-dim spacetime reflections. Moreover, Kuckert and Lorenzen (2007: 830) have demonstrated that the two representations of the Poincaré group obtained by these different methods are equivalent.

Poincaré Covariance from CGMA and CMG

The common goal of the approaches of Guido and Longo and Kuckert et al. is to construct a representation of the Poincaré group that acts on a net of local algebras. This is accomplished by assuming the net is invariant under a representation that implements a *subset* of Poincaré transformations. Guido and Longo start with a representation that implements (1 + 1)-dim Lorentz boosts on wedge regions by the modular operators associated with the wedges (the content of MC). Kuckert et al. start with a representation that implements (1 + 1)-dim PT transformations on wedge regions by the modular conjugations associated with the wedges (the content of MPCT). Thus in these approaches to the CPT and spin–statistics theorems, in replacing RLI with either MC or MPCT, one is replacing invariance under the restricted Lorentz group with invariance under a subgroup of the Poincaré group.

One might wonder if one can go further; in particular, if one can replace invariance under a subgroup of the Poincaré group with more general assumptions, derived ultimately from the modular data associated with a net of algebras. This question was answered in the affirmative in work initiated by Buchholz and Summers (1993).[24] This work first introduces a generalization of MPCT that it refers to as the CGMA. Buchholz et al. (2000: 477) state this condition in the following way: consider a family of wedge regions $\mathcal{W}$ in Minkowski spacetime, and a net $W \mapsto \mathfrak{R}(W)$ of von Neumann algebras defined on $\mathcal{W}$. CMGA requires the set $\{\mathfrak{R}(W)\}_{W\in\mathcal{W}}$ to be invariant under the adjoint action of the modular conjugations $\{\mathcal{J}_W\}_{W\in\mathcal{W}}$. In other words, for every W_1, $W_2 \in \mathcal{W}$, there is a $W_3 \in \mathcal{W}$ such that

$$\mathcal{J}_{W_1}\mathfrak{R}(W_2)\mathcal{J}_{W_1} = \mathfrak{R}(W_3). \qquad \text{(CGMA)}$$

Note that MPCT entails CGMA, but the latter is more general than the former, insofar as CGMA does not specify the form of the action of $\mathcal{J}_W$ on the wedge algebras. In particular, CGMA does not assume that the modular conjugations implement (1 + 1)-dim spacetime reflections on wedges.

One can also consider a generalization of MC referred to as the CMG, which requires the set $\{\mathfrak{R}(W)\}_{W\in\mathcal{W}}$ to be invariant under the adjoint action of the modular operators $\left\{\Delta_W^{it}\right\}_{W\in\mathcal{W}}$ (see, e.g., Buchholz et al., 2000: 532). Thus, for every $W_1, W_2 \in \mathcal{W}$ and any $t \in \mathbb{R}$ there exists a wedge $W_t \in \mathcal{W}$ such that,

$$\Delta_{W_1}^{it}\mathfrak{R}(W_2)\Delta_{W_1}^{-it} = \mathfrak{R}(W_t). \qquad \text{(CMG)}$$

Again, while MC entails CMG, the converse does not hold, insofar as CMG does not specify the form of the action of $\Delta_{W_0}^{it}$ on the wedge algebras. In particular, CMG does not assume that the modular operators implement $(1+1)$-dim Lorentz boosts on wedges.

Buchholz et al. (2000) now demonstrate that a representation of the Poincaré group can nevertheless still be constructed based on either CGMA or CMG, in conjunction with a few additional assumptions. In both cases one shows that the groups $\mathcal{J}$ and $\mathcal{K}$ generated by the set of modular conjugations and the set of modular operators, respectively, are homomorphic to transformation groups $\mathcal{T}$ and $\mathcal{U}$ on the wedges, that are isomorphic to P_+ and $\tilde{P}_+^{\uparrow}$, respectively.

To flesh out this program in slightly more detail, the additional assumptions can be summarized by the following conditions on the net $W \mapsto \mathfrak{R}(W)$ and the modular data $(\mathcal{J}_W, \Delta_W^{it})$.[25]

(a) *Formative Assumptions:*
- The map $W \mapsto \mathfrak{R}(W) \in \{\mathfrak{R}(W)\}_{W\in\mathcal{W}}$ *is an order-preserving bijection.*
- *If* $W_1 \cap W_1 \neq \varnothing$, *then* Ω *is cyclic and separating for* $\mathfrak{R}(W_1) \cap \mathfrak{R}(W_2)$. *Conversely, if* Ω *is cyclic and separating for* $\mathfrak{R}(W_1) \cap \mathfrak{R}(W_2)$, *then* $\bar{W}_1 \cap \bar{W}_2 \neq \varnothing$, *where the bar denotes closure.*

(b) *Transitivity of* $\mathcal{J}$*: The adjoint action of the modular conjugations* $\mathfrak{J}_W$, $W \in \mathcal{W}$, *acts transitively on* $\{\mathfrak{R}(W)\}_{W \in \mathcal{W}}$.

(c) *Transitivity of* $\mathcal{K}$*: The adjoint action of the modular operators* Δ^{it}_W, $W \in \mathcal{W}$, *acts transitively on* $\{\mathfrak{R}(W)\}_{W \in \mathcal{W}}$.

Transitivity of $\mathcal{J}$ requires that there exists a wedge $W_0 \in \mathcal{W}$ such that $\{\mathfrak{J}_W \mathfrak{R}(W_0) \mathfrak{J}_W\}_{W \in \mathcal{W}} = \{\mathfrak{R}(W)\}_{W \in \mathcal{W}}$, and similarly for transitivity of $\mathcal{K}$, with $\mathfrak{J}_W$ replaced by Δ^{it}_W. Conditions (b) and (c) entail that the wedge transformation groups $\mathcal{T}$ and $\mathcal{U}$ act transitively on the set of wedges $\mathcal{W}$.[26] Buchholz et al. (2000: 527, 535) now demonstrate the following entailments:

(I) [CGMA & (a) & (b) & (*net continuity*)] $\Rightarrow$ [(existence of $U(P_+)$ & MPCT]

(II) [CMG & (a) & (c)] $\Rightarrow$ [existence of $U(\tilde{P}^{\uparrow}_{+})$ & MC]

In entailment (I), *net continuity* is a technical condition that guarantees that the unitary representation U of the Poincaré group in the first entailment is continuous (Buchholz et al., 2000: 517), and MPCT is the statement $U(g_W) = \mathfrak{J}_W$, where g_W is the $\mathrm{P}_1\mathrm{T}$ transformation that reflects W about the x_1-axis (in other words, under the representation U, the modular conjugation $\mathfrak{J}_W$ implements $(1+1)$-dim spacetime reflections). Similarly, in entailment (II), MC is the statement that the modular operators associated with a wedge implement $(1+1)$-dim Lorentz boosts. These entailments show that the construction of a representation of the Poincaré group can go through without the assumptions of MPCT or MC. Entailments (I) and (II) also explicitly indicate that CGMA and CMG are weaker assumptions than MPCT and MC.

Buchholz et al. (2000: 536, 538) also establish the following entailments relating CGMA and CMG:

(III) [CMG (a) & (c) & (*algebraic causality*)] $\Rightarrow$ [CGMA & MPCT]

(IV) [CGMA & (a) & (b) & (c) & (*modular stability*)] $\Rightarrow$ MC

In entailment (IV), *modular stability* requires that the modular operators are contained in $\mathcal{J}$.[27]

Two observations can be made at this point. First, evidently, proofs of the spin–statistics and CPT theorems can now be constructed by replacing MC and MPCT in the proofs of Guido and Longo, and Kuckert et al., with CMG and CGMA (along with the pertinent additional assumptions) (Buchholz et al., 2000: 532). Second, in these proofs, the role of RLI is played by the explicit construction of a representation of the Poincaré group, now not from implementations of its subgroups (as in the cases of the proofs of Guido and Longo and Kuckert et al.), but from an implementation of the transformations between wedge regions in Minkowski spacetime.

2.2.2 On the Fundamentality of Constraints on Modular Data

The initial goal of algebraic approaches to the CPT and spin–statistics theorems is to obtain a representation of the Poincaré group that acts on a net of local algebras. We now have four methods of obtaining this goal:

(a) Assume the modular operators associated with wedge regions implement $(1 + 1)$-dim Lorentz boosts (MC).

(b) Assume the modular conjugations associated with wedge regions implement $(1 + 1)$-dim spacetime reflections (MPCT).

(c) Assume the modular conjugations associated with wedge regions implement wedge transformations (CGMA).

(d) Assume the modular operators associated with wedge regions implement wedge transformations (CMG).

Each of these assumptions is weaker than an assumption associated with RLI, as can be seen by the following considerations. First, if RLI translates into the algebraic context as the existence of a representation of the Poincaré group that acts on a net of local algebras, then certainly assumptions (a)–(d) are weaker than RLI. Moreover, if RLI retains the meaning it has in the Wightman axiomatic approach, then assumptions (a)–(d) are weaker than RLI in the following sense: if the local algebras $\mathfrak{R}(\mathcal{O})$ are generated by local finite-component Wightman fields that satisfy RLI (as well as SC and LC), then (a) and (b) are consequences of the Bisognano–Wichmann theorem, and (c) and (d) are then entailed by (a) and (b), respectively. Having thus established that assumptions (a)–(d) are weaker than RLI, the questions to be addressed in this section are, can they be independently motivated, and if so, can any one of them be identified as weaker than the others, i.e., which should we take to be more fundamental?

The Fundamentality of MC

Physically, Guido and Longo (1995: 520) suggest MC may be motivated by appeal to the Unruh effect.[28] This effect occurs when an observer, in constant acceleration with respect to the Minkowski vacuum, experiences the latter as a thermal state.[29] A standard explanation runs as follows: one can show that the restriction $\omega_0 \mid_{\mathfrak{R}(W)}$ of the Minkowski vacuum state ω_0 to the wedge W is a Kubo–Martin–Schwinger (KMS) state with respect to the group $\mathcal{K}$ generated by the modular operators associated with W (for a definition of the KMS condition, see Haag, 1996: 218). Now suppose KMS states are identified as thermalized states, i.e., equilibrium states at some finite temperature (for motivation, see Earman, 2011: 82–83). MC then entails that $\omega_0 \mid_{\mathfrak{R}(W)}$ is a thermalized state with respect to Lorentz boosts on W. Since the orbits of the latter are world lines of constant acceleration, the Unruh effect follows, i.e., an observer in constant acceleration experiences $\omega_0|_{\mathfrak{R}(W)}$ as a thermalized state. This modular theory derivation of the Unruh effect can be summarized in the following arguments:

1. The restriction $\omega_0|_{\mathfrak{R}(W)}$ is a KMS state with respect to $\mathcal{K}$.
2. KMS states are thermalized states (i.e., equilibrium states at some finite temperature).
3. (MC) $\mathcal{K}$ generates Lorentz boosts on W.

4. The restriction $\omega_0|_{\mathfrak{R}(W)}$ is a thermalized state with respect to Lorentz boosts on W.

1′. The restriction $\omega_0|_{\mathfrak{R}(W)}$ is a thermalized state with respect to Lorentz boosts on W.

2′. The orbits of Lorentz boosts on W define paths of constant acceleration with respect to W.

3′. (Unruh effect) An observer in constant acceleration with respect to W experiences the restriction as a thermalized state.

Guido and Longo now suggest,

> ... [T]he equivalence principle in relativity theory then allows an interpretation of the thermal outcome as a gravitational effect. On this basis Haag has proposed long ago to derive the Bisognano–Wichmann theorem [and thus MC]. (Guido and Longo, 1995: 520)

The suggestion, evidently, is that MC coupled with the equivalence principle entails that a gravitational field thermalizes the vacuum state of a von Neumann algebra of observables. The extent to which this constitutes a motivation for MC may initially depend on one's attitudes toward explanation and/or confirmation. For instance, to the extent that explanation requires derivation from first principles, MC might be claimed (in part) to explain the Unruh effect, insofar as MC appears as one of the premises in the modular theory derivation of the Unruh effect. Similarly, to the extent that confirmation requires derivation of evidence from hypothesis, evidence for the Unruh effect might be claimed to be evidence for MC. However, these considerations don't stand up to further scrutiny. Apart from questions concerning the feasibility of applying the equivalence principle in the context of flat Minkowski spacetime, one may also question the cogency of the above modular theory derivation of the Unruh effect. Earman (2011: 87–88) raises the following concerns. First, the relevant KMS state is the restriction $\omega_0|_{\mathfrak{R}(W)}$ of the Minkowski vacuum state to the wedge algebra $\mathfrak{R}(W)$. This is the vacuum state experienced by an observer in perpetual constant acceleration who has access only to $\mathfrak{R}(W)$. On the other hand, an observer who maintains constant acceleration for any finite stretch of proper time τ, no matter how long, but is unaccelerated either at $\tau = +\infty$ or $\tau = -\infty$, will have access to the full quasi-local algebra $\mathfrak{R}$ and the corresponding Minkowski vacuum state ω_0, for which the KMS result does not hold. This suggests that whether or not an observer experiences the Unruh effect cannot be determined by facts about any finite portion of her history, and "[t]his makes it mysterious how to mesh the deliverances of modular theory with the registrations of laboratory instruments" (Earman, 2011: 88). Second, to argue that the KMS states associated with the modular group of $\mathfrak{R}(W)$ are thermal states depends on an analogy between KMS states and Gibbs states in quantum statistical mechanics. In the first instance, this requires an assumption that the modular group parameter can be interpreted as inverse temperature. In some contexts, the justification for this assumption is based on systems characterized by KMS states and obtained by taking appropriate thermodynamic limits of ordinary thermodynamic systems. But no such limiting procedures are associated with the restriction of the Minkowski vacuum state to $\mathfrak{R}(W)$. Moreover, even if this assumption is accepted, one is still faced with the task of explaining, in the context of RQFT, how thermodynamic effects physically arise for accelerating observers. Assumedly such an explanation would require an account of how the vacuum state couples to accelerating observers, an account that modular theory by itself does not furnish.

The upshot of the above discussion is that it is still an open question as to whether MC is a physically reasonable fundamental assumption.

The Fundamentality of MPCT

Kuckert (2007: 227) observes that Guido and Longo (1995) derive MPCT in the process of constructing their representation and concludes that theirs is the stronger assumption:

> On the other hand, Guido and Longo make the stronger symmetry assumption (as their derivation of modular P_1CT-symmetry from the Unruh effect [i.e., MC] shows)... The Unruh effect [i.e., MC] itself already entails full Lorentz symmetry, whereas partial P_1CT-symmetry merely entails rotational symmetry (Kuckert, 2007: 227).

Note that it is not entirely clear how MC should be associated with "full Lorentz symmetry," since it only assumes the implementation of $(1+1)$-dim Lorentz *boosts* on wedges. Moreover, MPCT assumes the implementation of $(1 + 1)$-dim spacetime reflections, hence it seems to entail more than just "rotational symmetry." On the other hand, Kuckert's claim that MC is the "stronger symmetry assumption" is born out in the analysis of Buchholz et al. (2000). Entailment (I) of Section 2.3.1 indicates that CGMA implies MPCT (with the appropriate additional assumptions). Moreover, Buchholz et al. (2000) provide examples of Poincaré covariant nets that satisfy CGMA (and the same appropriate additional assumptions) but violate MC, either with or without the SC; they conclude: "It is therefore clear that the assumption of modular covariance is more restrictive than CGMA, even when the spectrum condition is posited" (Buchholz et al., 2000: 539).

Recall that Guido and Longo suggest that MC can be independently motivated by an (ultimately suspect) appeal to the Uhruh effect. Kuckert does not provide a similar independent motivation for MPCT. In any event, in light of the analysis of Buchholz et al. (2000), the more relevant question is whether there is independent motivation for either CGMA or CMG.

The Fundamentality of CGMA and CMG

From Section 2.2.1, entailment (I) indicates that CGMA (with the appropriate additional assumptions) is weaker than MPCT, whereas entailment (II) indicates that CMG (with the appropriate additional assumptions) is weaker than MC. Moreover, entailments (III) and (IV) indicate that CGMA is weaker than CGM. This might suggest that CGMA is a more fundamental constraint than CGM. Buchholz et al. (2000) further suggest that CGMA is a more intrinsically spatiotemporal condition than CGM. With respect to the goal of constructing a representation of the Poincaré group, they observe,

> ... since the modular involutions [i.e., conjugations] depend only on the characteristic cones of the pairs $(\mathfrak{R}(W), \Omega)$, it would seem that they are more likely to encode some intrinsic information about the representation as opposed to the modular unitaries [i.e., operators], which are strongly state-dependent. (Buchholz et al., 2000: 542)

In other words, a pair $(\mathfrak{R}(W), \Omega)$ of a local algebra and a cyclic and separating vacuum vector contains both spatiotemporal and algebraic information. The corresponding modular conjugation encodes the former, to the extent that it depends only on the light-cone structure associated with the spacetime region W, whereas the corresponding modular operators encode the latter. The association of the modular conjugation with light-cone structure is based on Buchholz et al.'s (2000) proof of entailment (I) in Section 2.2.1. The sense in which the modular operators are "strongly state-dependent" and encode algebraic information may be fleshed out in terms of the properties of the modular automorphism group $\mathcal{K}$ they generate: $\mathcal{K}$ encodes information about $\mathfrak{R}$ ($\mathcal{K}$ is an inner automorphism on $\mathfrak{R}$ if and only if $\mathfrak{R}$ is semifinite), and information about Ω (the state ω induced by Ω is invariant under $\mathcal{K}$, and is tracial if and only if $\mathcal{K}$ is trivial) (Summers, 2006: 252).

These considerations make it reasonable to identify CGMA as a more fundamental modular constraint than MC, MPCT, and CMG when it comes to constructing a representation of the Poincaré group. The question remains as to whether there are independent physically reasonable motivations for adopting CGMA. Here I can only report on recent work that might provide such motivation. Note that while CGMA and CMG do not assume that the adjoint actions of the modular objects implement spacetime transformations (boosts and PT reflections), they still assume an underlying spacetime, if only to define the wedge regions. In recent work, Summers and colleagues have embarked on a program to reconstruct the metrical structure of a background spacetime from the algebraic properties of the modular data of a net of algebras indexed, not by wedges in Minkowski spacetime, but by abstract "laboratories" (Summers and White, 2003: 204). In particular the task is to reconstruct the isometry group of a spacetime as the group $\mathcal{J}$ generated by the modular conjugations of such a net, constrained by CGMA and an additional set of purely algebraic axioms.[30] This program has been carried out for 3-dim de Sitter spacetime, and Minkowski spacetime in three and four dimensions (Summers, 2011: 335). Thus for these examples, one could argue that CGMA is a physically well-motivated assumption in the sense that it (together with the additional axioms) encodes the spatiotemporal information associated with a net of local algebras, with no need to specify this information independently of the net.

2.3 Relativity and CPT Invariance

Sections 2.1 and 2.2 have told us that relativity, in the form of RLI, is neither necessary nor sufficient to derive the SSC and CPT invariance. An influential claim made by Greenberg (2002) appears to suggest otherwise. Greenberg claims, "If CPT invariance is violated in an interacting quantum field theory, then that theory also violates Lorentz invariance" (Greenberg, 2002: 1). This section investigates the nature of this claim.

2.3.1 An Influential But Puzzling Claim

Greenberg's claim is both influential and puzzling. In the physics literature it is influential since it suggests a test of violations of Lorentz invariance by experiments that measure CPT violation. For instance, in a review of tests of Lorentz invariance, Liberati states:

> Note that Lorentz violation does not imply CPT violation for local EFTs, while CPT violation does imply Lorentz violation in local EFTs. (Liberati, 2013: 12)

A reference to Greenberg follows. Greenberg is also cited by Sozzi in a text on CP violation:

> In all proofs of the CPT theorem Lorentz symmetry is the basic hypothesis, and indeed a theorem states that if CPT symmetry is violated then Lorentz symmetry must be violated, too... (Sozzi, 2008: 198).

Finally, citations to Greenberg can also be found in the contributions to the proceedings of a conference on CPT invariance. Berger is a typical example:

> In realistic field theories, CPT violation is always accompanied by Lorentz violation, but not vice versa. (Berger, 2011: 180)

Greenberg's claim has also been influential in the philosophy of physics literature, but perhaps more indirectly. While none of the following authors directly cites Greenberg, their claims seem to be a reflection of his. Hagar for instance states:

> ... the CPT theorem... says that violations of CPT symmetry imply violations of Lorentz invariance, but not vice versa. (Hagar, 2009: 261)

Another example is an article on a geometric understanding of the CPT theorem by Greaves, who asks,

> How can it come about that one symmetry (e.g., Lorentz invariance) entails another (e.g., CPT) *at all*? (Greaves, 2010: 28)

Finally, in a review of the CPT theorem, Arntzenius states:

> The CPT theorem says that any (restricted) Lorentz invariant quantum field theory must also be invariant under the combined operation of [CPT]. (Arntzenius, 2011: 633)

Greenberg's claim is puzzling for two reasons. First, in both the purist and pragmatist proofs of the CPT theorem reviewed in Chapter 1, more than just Lorentz invariance was needed to derive CPT invariance. Additional assumptions include the SC, WLC, and CD, among others. Recall that there are multiple ways of formulating the CPT theorem that appeal to distinct sets of assumptions; and one of these ways (the algebraic approach) doesn't even require Lorentz invariance. The second reason Greenberg's claim is puzzling is the fact that purist and pragmatist proofs of the CPT theorem hold for *non-interacting*, and at most, *unrealistic interacting* states. In order to extend the proofs to realistic interacting fields, both the purist and the pragmatist need to confront the *existence problem* in Section 1.4.3. This problem intimately depends on the type of interaction, and this suggests that a demonstration of CPT invariance for realistic interacting fields may have to be done on a case by case basis. Thus, on the surface, Greenberg's claim seems to both simplify the assumptions needed to derive CPT invariance, and address the issue of the extent of its applicability in one fell swoop. The questions to be addressed below are, in the first instance, "Does Greenberg's claim amount to a purist or a pragmatist attempt to extend CPT invariance to realistic interacting RQFTs?" and, in the second instance, "Does it succeed?;" more precisely, "Does it address the existence problem?"

2.3.2 Greenberg's Argument

Greenberg begins his argument by asserting that "To calculate the S-matrix we need τ functions..." (Greenberg, 2002: 1). Recall from Section 1.4.1 that a τ-function is a vacuum expectation value of a time-ordered product of fields;

in other words, a τ-function is a time-ordered Wightman function. In general, an n-point τ-function is given by the following expression,

$$\tau^{(n)}(x_1, \ldots, x_n) \equiv \sum_p \theta\left(t_{p_1} - t_{p_2}\right) \ldots \theta\left(t_{p_{n-1}} - t_{p_n}\right) W^{(n)}\left(x_{p_1}, \ldots, x_{p_n}\right) \tag{2.5}$$

where the product of Heaviside functions $\theta(t_{p_1} - t_{p_2}) \ldots \theta(t_{p_{n-1}} - t_{p_n})$ enforces the time ordering $t_{p_1} > \ldots > t_{p_n}$ on the Wightman function $W^{(n)}$, and the sum is over all permutations of the indices. An example of a τ-function is the Feynman propagator $\Delta_F(x_1, x_2)$, which is given by the 2-point τ-function:

$$\tau^{(2)}(x_1, x_2) = \theta(t_1 - t_2)\,\langle 0\,|\phi(x_1)\phi(x_2)|\,0\rangle + \theta(t_2 - t_1)\,\langle 0\,|\phi(x_2)\phi(x_1)|\,0\rangle\,.$$

This is typically described as representing the propagation of a particle from x_1 to x_2, or x_2 to x_1, depending on their time order. Recall, again, from Section 1.4.1 that the significance of τ-functions is that they figure into an algorithm for calculating S-matrix elements. This is summarized in Table 2.2. One first reduces S-matrix elements to interacting τ-functions *via* the LSZ reduction formula (1.3) of Section 1.4.1; call this step (a). Pragmatist approaches to RQFTs then typically calculate interacting τ-functions by two additional steps:

(b) Express interacting τ-functions in terms of non-interacting τ-functions by means of the Gell-Mann–Low formula (1.4) of Section 1.4.1.

(c) Express non-interacting τ-functions as sums of products of Feynman propagators. This step makes use of Wick's theorem and can be graphically represented using Feynman diagrams.

Purist approaches to RQFTs may employ step (a) and use the LSZ reduction formula to obtain S-matrix elements, but such approaches replace the pragmatist's steps (b) and (c) with

(b′) Obtain interacting τ-functions directly from an appropriate interacting model of the relevant axioms.

Table 2.2 *Pragmatist and purist algorithms for interacting RQFTs.*

(a) (LSZ): S-matrix elements obtained from interacting $\tau^{(n)}$s.	
Pragmatist subroutine:	*Purist subroutine:*
(b) (Gell-Mann–Low): Interacting $\tau^{(n)}$s obtained from free $\tau^{(n)}$s.	(b′) Interacting $\tau^{(n)}$s obtained from interacting model of axioms.
(c) (Wick): Free $\tau^{(n)}$ expressed as sum of products of $\Delta_F(x_1, x_2)$.	

Note that the *existence problem* crops up in the pragmatist step (b) insofar as the Gell-Mann–Low formula involves a perturbative power series expansion that, for realistic interacting theories, does not satisfy any of the Pragmatist's existence criteria (Section 1.4.3). And, of course, the *existence problem* crops up in the purist step (b′) insofar as there are currently no realistic interacting models of the relevant sets of axioms.

The role that τ-functions play in an interacting RQFT suggests to Greenberg that Lorentz invariance be understood in terms of τ-functions in the following way:

> A quantum field theory is Lorentz covariant in cone if vacuum elements of unordered products of fields (Wightman functions) are covariant... A quantum field theory is covariant out of cone if vacuum matrix elements of time-ordered products (τ functions) are covariant... We require covariance of a quantum field theory both in and out of cone as the condition for Lorentz invariance of the theory. (Greenberg, 2002: 1)

In addition, Greenberg assumes that a theory's Wightman functions are (restricted) Lorentz invariant (RLI), and satisfy the SC.[31] Greenberg now argues that, at a Jost point, if a τ-function and its corresponding Wightman function satisfy RLI, then the Wightman function satisfies WLC. This follows from entailment (B) in Section 2.1.1, insofar as LC entails WLC. Greenberg (2002: 1) then argues that, since WLC of Wightman functions is necessary and sufficient for CPT invariance (given the assumption that the Wightman functions satisfy RLI and SC),

> ... if CPT invariance does not hold for this matrix element [viz., Wightman function], then the τ function is not Lorentz invariant and the theory is not Lorentz invariant.

The argument so far can be reconstructed in the following way:

1. An RQFT is Lorentz invariant *iff* both its Wightman functions and its τ-functions are RLI.
2. An RQFT has Wightman functions that satisfy RLI and SC.
3. If a τ-function is RLI at a Jost point, then, if its corresponding Wightman function satisfies RLI, then it satisfies WLC at that Jost point.
4. If an RQFT has Wightman functions that satisfy RLI, SC and, in addition, satisfy WLC at a Jost point, then it is CPT invariant.

(C) A violation of CPT invariance in an RQFT entails a violation of Lorentz invariance.

Premise (4) is Jost's axiomatic CPT theorem (Section 1.2.4). By premises (4) and (2), a violation of CPT invariance entails that a theory's Wightman functions fail to satisfy WLC at Jost points. By premise (3), this entails that the theory's

τ-functions fail to satisfy RLI at Jost points. By premise (1), this entails the theory is not Lorentz invariant, and thus the conclusion (C) follows.

Note that no mention of an *interacting* theory has been made at this point. Indeed, Greenberg now claims,

> This argument does not apply to a non-interacting theory for which τ functions need not be considered. Thus we have demonstrated the main result of this paper. *If CPT invariance is violated in an interacting quantum field theory, then that theory also violates Lorentz invariance.* (Greenberg, 2002: 1)

This suggests that for Greenberg, an interacting RQFT is characterized by both Wightman and τ-functions, whereas a non-interacting RQFT is characterized by Wightman functions alone. On first blush, this seems plausible, given the algorithm in Table 2.1 for calculating S-matrix elements from τ-functions. Thus if an interacting RQFT is in the business of calculating S-matrix elements, then τ-functions can play an important role. But this line of reasoning breaks down if there are other methods for calculating S-matrix elements that do not rely on τ-functions, and, moreover, if there are other testable predictions of interacting RQFTs that can be derived without the use of τ-functions. I will bracket these concerns off for the time being, since there seems to be a more fundamental difficulty with Greenberg's argument.

The implication that Greenberg draws from the significance of τ-functions to interacting RQFTs seems to be that, in the reconstructed argument above, "RQFT" can be replaced with "interacting RQFT" in premises (1) and (2), with the result that the conclusion can now explicitly refer to interacting RQFTs:

1′. An interacting RQFT is Lorentz invariant *iff* both its Wightman functions and its τ-functions are RLI.

2′. An interacting RQFT has Wightman functions that satisfy RLI and SC.

3. If a τ-function is RLI at a Jost point, then, if its corresponding Wightman function satisfies RLI, then it satisfies WLC at that Jost point.

4. If an RQFT has Wightman functions that satisfy RLI, SC and, in addition, satisfy WLC at a Jost point, then it is CPT invariant.

(C′) A violation of CPT invariance in an interacting RQFT entails a violation of Lorentz invariance.

Premise (4), the statement of the axiomatic CPT theorem, remains unaltered. Again, the question this section seeks to answer is whether Greenberg successfully extends CPT invariance to realistic interacting RQFTs. The modified argument is still valid: by premises (4) and (2′), a violation of CPT invariance in an interacting RQFT entails that its Wightman functions violate WLC at Jost points. By premise (3), this entails that the interacting theory's τ-functions fail to satisfy RLI at Jost points, and by premise (1′) this entails that the interacting theory is not Lorentz invariant; hence the conclusion (C′).

However, it now becomes apparent that, even if we allow premise (1′) and the implied significance of τ-functions to interacting RQFTs, Greenberg's argument, extended to *realistic* interacting RQFTs, breaks down under either a pragmatist or a purist interpretation. Under a purist interpretation, Greenberg's modified argument fails for *realistic* interacting RQFTs insofar as, for such theories we have reason to reject premise (2′). In particular, there are no known models of the Wightman axioms (or other purist axioms) that describe realistic interacting states that satisfy SC.[32] Since SC is an axiom in purist approaches, failure of SC for realistic interacting states is an instance of the *existence problem* for purists.

Does Greenberg's argument fare any better as a *pragmatist* attempt to extend CPT invariance to realistic interacting RQFTs? Note first that pragmatists have a bit more to say about premise (1′) than purists. In Weinberg's approach, for instance, we have the following implications (Weinberg, 1995: 144–145):

$$\begin{aligned}(\mathfrak{H}_{int}(x) \text{ satisfies RLI and } \textit{causality}) \quad &\Rightarrow \quad (\text{time-ordered products of } \mathfrak{H}_{int}(x) \\ &\qquad \text{are RLI}) \\ &\Rightarrow \quad (\text{RLI of } S\text{-matrix})\end{aligned}$$

where $\mathfrak{H}_{int}(x)$ is the theory's interaction Hamiltonian density. The first line is a special case of entailment (A) in Section 2.1.1. The second line follows since if time-ordered products of $\mathfrak{H}_{int}(x)$ are RLI, then so is the S-matrix in the form of equation (1.2) in Section 1.2.2, since all other quantities in equation (1.2) are manifestly RLI.[33]

Thus if an RQFT is identified with its S-matrix, then a violation of RLI of its τ-functions does not necessarily entail a violation of RLI of the theory. On the other hand, if an RQFT is identified with its Hamiltonian density, then a violation of RLI of its τ-functions entails either the theory violates RLI, or it is nonlocal (in the sense that its Hamiltonian density does not commute at spacelike separations).

Now recall that pragmatists justify the existence of realistic interacting τ-functions, not by providing provisos concerning the possibility of constructing realistic interacting models of a set of axioms, but rather by employing the Gell-Mann–Low formula. However, this still confronts them with the *existence problem*, but now in its pragmatist form. This problem makes its explicit appearance for a pragmatist in premise (2′) insofar as realistic interacting Wightman functions obtained from the Gell-Mann–Low formula do not satisfy the SC.

The upshot of this discussion is that, considered as either a purist or a pragmatist attempt to extend CPT invariance to realistic interacting fields, the soundness of Greenberg's argument stands or falls with the *existence problem. If* we could solve this problem and construct a realistic interacting model of a relevant set of axioms (or if the pragmatist's *convergence problem* could be solved), then Greenberg's argument would, perhaps, be sound (modulo the concerns with premise 1). As it stands, however, to accept Greenberg's argument is to brush the *existence problem* under the rug. While his demonstration is insightful in uncovering connections between Lorentz invariance and CPT invariance in abstract objects like

τ-functions and Wightman functions, both purists and pragmatists should be hesitant in extending these observations to realistic interacting RQFTs, at least under the current state of affairs.

2.4 Summary

What is the role of relativity in deriving the spin–statistics connection (SSC) and CPT invariance? The answer to this question depends first on what one means by relativity. If relativity is understood as invariance under the restricted Lorentz group (RLI), and if this invariance is implemented by a representation of the covering of the restricted Lorentz group on our space of states, then, in the first instance,

(a) The role of RLI in derivations of SSC is that it is a way of encoding spin: the spin of a state can be encoded in the representation of the covering of the restricted Lorentz group that leaves the state invariant.

(b) The role of RLI in derivations of CPT invariance is that it is a way of encoding PT invariance.

However, if we adopt RLI in the pursuit of (a) and (b), we still need additional assumptions in order to derive SSC and CPT invariance. In particular, we need a kinematical locality assumption that takes the general form of a prohibition on superluminal signaling. For the standard derivations reviewed in Chapter 1, this locality assumption comes in one of three forms: LC, CD, or *causality*.

We've also seen in Section 2.2 that RLI is not necessary in order to encode spin, nor in order to encode PT invariance. There are weaker assumptions that allow us to obtain representations of spin and PT transformations than assuming our theory is invariant under the restricted Lorentz group (and that still allow us to derive SSC and CPT invariance). Namely, we can employ the algebraic approach and adopt any one of MC, MPCT, CGMA, or CMG to replace RLI.

This may seem slightly disingenuous: in the algebraic approach, the initial goal was to construct a representation of the Poincaré group that implemented (a) and (b) above. Thus, while we did not assume RLI initially, it was derived in the process of deriving SSC and CPT invariance. This leaves open the question of whether SSC and CPT invariance can be derived in non-relativistic theories. This question will be addressed in Chapters 3 and 4.

NOTES

1. Strictly speaking, the axiomatic approach adopts what Section 1.1.2 called the *statistics–locality connection* (StLC) which is a stronger version of LC. See Sections 2.1.1 and 2.1.4.

2. The quotes from Haag, Strocchi, and Streater and Wightman are intended as justifications for an axiomatic version of CD that applies to Wightman functions, but that entails CD of the S-matrix. See Section 2.1.3.
3. Earman and Valente's intent is to articulate a notion of "relativistic causality" in the algebraic approach. They ultimately associate this notion with a particular algebraic formulation of no superluminal signaling.
4. Similarly a sense of (non-) locality associated with the notion of holism can be ignored.
5. The kinematical aspects of a theory involve the specification of the structure of a theory's state space (the sorts of things the theory identifies as possible states of the physical systems the theory purports to describe, and the constraints imposed on such states). The dynamical aspect of a theory involves the specification of how the possible states of the theory evolve in time.
6. "At best," since it is not that clear that the argument that equates *algebraic causality* with NSS can be run in the other approaches, and since, in any event, adopting NSS as a constraint on a theory requires complicated issues involving the definition of a signal (Weinstein, 2006).
7. Geroch (2011) argues, convincingly, for the compatibility of RLI on the one hand, and superluminal propagation, on the other.
8. Note that RLI for Wightman functions entails RLI for the corresponding fields, provided the vacuum state is RLI.
9. To see this, let $\Delta t = t_1 - t_2$ encode the time ordering of two events x_1, x_2 separated by a spacetime interval Δs, i.e., Δt is positive (resp. negative) just when x_1 occurs after (resp. before) x_2. Under a Lorentz transformation, Δt transforms according to $c\Delta t' = \gamma\left[c\Delta t - (v/c^2)\Delta x\right]$, where $\gamma = \left(1 - v^2/c^2\right)^{-1/2}$, and Δx is the spatial part of Δs. If Δs is *timelike* or *lightlike*, then $c\Delta t \geq \Delta x$; thus $c\Delta t > \left(v/c^2\right)\Delta x$, since $v/c^2 < 1$. If Δt is positive (*resp.* negative), then $\Delta t'$ has to be positive (resp. negative), too. Thus time-ordering is preserved for timelike and lightlike intervals. If Δs is *spacelike*, then $c\Delta t < \Delta x$; and if Δt is positive, then $\Delta t'$ can be negative, as long as $v/c^2 > c\Delta t/\Delta x$. Thus time-ordering is not necessarily preserved for spacelike intervals.
10. This is the argument that underwrites the sufficient condition for assumption (ii) in Weinberg's approach to the spin–statistics and CPT theorems (Section 1.2.2).
11. For anti-commuting fields, $T\left\{a_p(t)a_q(t')\right\} = -T\left\{a_q(t')a_p(t)\right\}$, "[t]hus we cannot just define time ordering as 'take all the operators and put them in time order', or else this equation would imply the time-ordered product must vanish;" so for anti-commuting fields, time ordering must be defined by $T\{\psi(x)\chi(y)\} = \psi(x)\chi(y)\theta(x_0 - y_0) - \chi(y)\psi(x)\theta(y_0 - x_0)$ (Schwarz, 2013: 212).
12. Greenberg notes that the Wightman reconstruction theorem is the reason for restricting the proof of entailment (B) to Jost points. This allows τ-functions to be equated with Wightman functions, which then allows access to the

Wightman reconstruction theorem. On the other hand, Greenberg (2006a: 2) indicates that entailment (B) holds not just for τ-functions, but also for "other Greens functions." In particular, it holds for *retarded* functions, for which there is a reconstruction theorem (Glaser et al., 1957). This suggests a variant of entailment (B) that replaces τ-functions with retarded functions and eliminates reference to Wightman functions.

13. Note that Glaser et al. (1957) derive a version of the LSZ reduction formula (equation (1.3)) that employs retarded functions.
14. Greenberg (1998: 145) also associates the "spin–statistics" theorem with Weinberg's approach: "[Fierz and Pauli] used locality of observables as the crucial condition for integer-spin particles and positivity of the energy as the crucial condition for the odd half-integer case. Weinberg showed that one can use the locality of observables for both cases if one requires positive-frequency modes to be associated with annihilation operators and negative-frequency modes to be associated with creation operators." However, for Weinberg, "locality of observables" (i.e., commutativity at spacelike separated distances) is only imposed on the interacting Hamiltonian density $\mathfrak{H}_{int}(x)$ and only to formally secure Lorentz invariance of the S-matrix (see Section 1.2.2). For Weinberg, "causality" as applied to observable quantities other than the S-matrix is explicitly renounced, as reflected in the quote in footnote 22 of Chapter 1.
15. Greenberg (1998: 145) associates the following assumptions with what he calls the spin–statistics theorem: "(1) that the space of states is a Hilbert space, i.e., the metric is positive-definite, (2) the fields smeared with test functions in the Schwarz space S have a common dense domain in the Hilbert space, (3) the fields transform under unitary representation of the restricted homogeneous Lorentz group, (4) the spectrum of states contains a unique vacuum and all other states have positive energy and positive mass, and (5) the bilinear observables constructed from the (free) asymptotic fields commute at spacelike separation..." The spin–locality theorem "... replaced assumption (5) of the spin–statistics theorem by (5′) that the fields either commute... or anticommute... at spacelike separation" (Greenberg, 1998: 146).
16. Entailment A1′ follows from entailment (ii′) in the Appendix.
17. Entailment C′ follows from combining entailments (b) and (c) in Section 1.2.3.
18. The Appendix suggests that Greenberg's assertion that the Luders–Zumino and Burgoyne proof has nothing to do with statistics fails to recognize the (perhaps implicit) assumption StLC. This assumption conjoined with the assumption of the wrong spin–statistics connection (NSSC), entails NSpLC, which is then key to the axiomatic proof that the fields vanish.
19. In equation (2.1), $\int dq_i$ represents a sum over spins and an integration over $d_3\mathbf{p}_i$.
20. See, e.g., Duncan (2012: 148). The proof requires the general result that any Fock space operator can be expressed as a sum of products of creation and

annihilation operators. One then considers a general $N'M'$ element of H, so expressed. It will decompose into a sum of disconnected elements and a fully connected element. For elements with $N' < N$ and/or $M' < M$, there are not enough $a^\dagger$, a operators to affect all N' particles in the initial state and/or all M' particles in the final state. Such elements thus contribute only to the disconnected components. For elements with $N' > N$ and/or $M' > M$, there are too many $a^\dagger$, a operators; some will eventually end up annihilating the vacuum. Hence the only part of H to contribute to the fully connected matrix element is h_{NM}.

21. $\{\Lambda_W(t), t \in \mathbb{R}\}$ is a subgroup of the restricted Poincaré group $P_+^\uparrow$ that acts on the coordinates (x_0, x_1, x_2, x_3) via the matrices

$$\begin{pmatrix} \cosh t & \sinh t & 0 & 0 \\ \sinh t & \cosh t & 0 & 0 \\ 0 & 0 & 1 & 0 \\ 0 & 0 & 0 & 1 \end{pmatrix}.$$

22. For a field of arbitrary spin, relation (c) is given by $\mathcal{J}_W = Z\Theta R_W$, where $Z = (I + i\Gamma)/(1+i)$ is a unitary operator on $\mathcal{H}$ defined in terms of the statistics operator Γ (Bisognano and Wichmann, 1976: 310).
23. These constructions are based on the facts that "each rotation can be obtained by combining two reflections by two-dimensional subspaces, and each restricted Lorentz transformation can be obtained by combining two reflections by two-dimensional *spacelike* subspaces" (Kuckert, 2007: 226). In this approach, the CPT operator is given by the composition $\mathcal{J}_a\mathcal{J}_b\mathcal{J}_c$ of three modular conjugations associated with "mutually orthogonal" wedges. Buchholz et al. (2000) and Buchholz and Summers (2004) earlier produced the construction for $\tilde{L}_+^\uparrow$.
24. This work includes Buchholz and Summers (1993), Buchholz et al. (1999), Buchholz et al. (2000), and Buchholz and Summers (2004). A summary is provided by Summers (2011).
25. Buchholz et al. (2000: 485, 532) redefine CGMA to include condition (a), and define CMG to be the conjunction of CMG as stated in the text above and condition (a). Other authors define CGMA in slightly different ways (see, e.g., Buchholz and Summers, 2004: 637–638; Summers, 2011: 331). However, the "core of this condition" (Buchholz and Summers, 2004: 638) is the invariance property denoted by CGMA in the text above.
26. Transitivity is required to guarantee that the transformation groups $\mathcal{T}$ and $\mathcal{U}$ are isomorphic to P_+ and $\tilde{P}_+^\uparrow$, respectively, as opposed to subgroups of the latter (Buchholz et al., 2000: 506).
27. One can show that the conjunction of CGMA, *modular stability*, condition (c), and *net continuity* entails the (negative or positive) *spectrum condition* (Buchholz et al., 2000: 529).

28. In fact, Kuckert views MC as a statement of the Unruh effect (Kuckert, 2005: 85; Kuckert 2007: 227; Kuckert and Lorenzen, 2007: 830).
29. This is typically interpreted as a thermalized multi-particle state, although Earman (2011) and Arageorgis et al. (2003) argue that this unjustified.
30. CGMA guarantees that the set $\{\mathfrak{J}_i\}$ of modular conjugations, where i is an element of an abstract set of labels, forms an "invariant generating set" for the group $\mathcal{J}$ (this means that $\mathcal{J}$ is the smallest group that contains $\{\mathfrak{J}_i\}$, and that for all $\mathfrak{J} \in \mathcal{J}, \mathfrak{J}\{\mathfrak{J}_i\}\mathfrak{J} \subset \{\mathfrak{J}_i\}$). The strategy is then to appropriate the techniques of "absolute geometry," which seeks to construct metric spaces from a group of motions characterized by an invariant generating set constrained by appropriate axioms (Summers, 2011: 333).
31. "We assume in cone Lorentz covariance in this paper" (Greenberg, 2002: 1). "We assume the spectrum of energy and momentum of the theory lies in or on the plus light cone" (Greenberg, 2002: 4).
32. The *spectrum condition* is essential to establish the analyticity of (complex) Wightman functions. Dütsch and Gracia-Bondía (2012: 429) observe that "... to the best of our knowledge, for non-trivial realistic models one cannot ascertain analyticity of Wightman-like functions; hence the argument a la Jost in [Greenberg, 2002] flounders."
33. This argument is based on the perturbative expression (1.2) for the S-matrix. Weinberg (1995: 119–120) also offers a *non-perturbative* proof that RLI and LC of the Hamiltonian density entails RLI of the S-matrix.

3

CPT Invariance and the Spin–Statistics Connection in Non-Relativistic Quantum Field Theories

This chapter addresses the question, "Why are CPT invariance and the spin–statistics connection (SSC) not derivable properties in non-relativistic quantum field theories (NQFTs)?" Recall from the introduction that an NQFT is a non-relativistic quantum theory with an infinite number of degrees of freedom. These degrees of freedom typically are interpreted as describing the values that a field can take at all points in spacetime, hence the term "field theory;" however, an NQFT can also be taken to describe a non-relativistic quantum many-body system composed of an infinite number of particles. NQFTs should be made distinct from theories that describe non-relativistic quantum systems with finite degrees of freedom, which are referred to in the Introduction as non-relativistic quantum mechanics (NQM) and which will be considered in Chapter 4.

Chapter 2 indicated that a simple appeal to the failure of relativity in NQFTs is not enough to explain why CPT invariance and the SSC are not essential properties in these theories. Simply put, in RQFTs, relativity (in the form of restricted Lorentz invariance) is neither necessary nor sufficient to derive these properties. The strategy in this chapter will be to formulate NQFTs within each of the approaches to RQFTs reviewed in Chapter 1, and then assess the failure of the CPT and spin–statistics theorems within each approach. The key idea is to take the differences in these approaches seriously: an explanation of why the CPT and spin–statistics theorems fail in *axiomatic* NQFTs, for instance, based on an appeal to the proofs of these theorems in *Weinberg's* approach, say, will be unsatisfying at best, and misleading at worst, given the conceptual and foundational differences between these approaches.

An essential part of this analysis will involve working out the intertheoretic relation between a given formulation of RQFT and its NQFT analog. Moreover, in three of these four formulations, we will see that, while the violation of restricted Lorentz invariance (RLI) in NQFTs plays a primary role in the failure of the CPT

CPT Invariance and the Spin–Statistics Connection. Jonathan Bain.

and spin–statistics theorems, it is not enough to *just* say this violation explains these failures: we will see that the explicit role that the violation of RLI plays varies between approaches. Thus an adequate explanation of why CPT invariance and the SSC are not derivable in NQFTs will in part have to involve identifying *how* the violation of RLI in a given approach blocks the formulation of a CPT and a spin–statistics theorem.

Section 3.1 sets the stage by describing the difference between RQFTs and NQFTs in terms of spacetime structure. Section 3.2 then considers how to formulate NQFTs in each of the approaches reviewed in Section 1.2, namely, the axiomatic approach, Weinberg's approach, the Lagrangian approach, and the algebraic approach:

(a) Section 3.2.1 considers the axiomatic formulation of Galilei-invariant quantum field theories (GQFTs) given by Lévy-Leblond (1967) and compares it with the Wightman axiomatic formulation of RQFTs. Lévy-Leblond (1967: 165) views the failure of (relativistic) *local commutativity* (LC) in GQFTs as the reason why the CPT and spin–statistics theorems fail. I will suggest that this is based on an appeal to Weinberg's pragmatist proofs, and thus, perhaps, is not appropriate for a purist approach. I will argue, instead, that a more appropriate explanation of why the CPT and spin–statistics theorems fail in axiomatic NQFTs should be based on considerations internal to the axiomatic approach. Such considerations indicate that the reason for the failure of these theorems in axiomatic NQFTs is twofold: first, the complex versions of classical spacetime symmetry groups (that underwrite NQFTs) do not have a component connected to the identity that contains the PT transformation, as the complex Lorentz group does; and second, a key property of the RQFT vacuum state (the "separating property") fails for NQFT vacuum states.

(b) Section 3.2.2 considers how Weinberg's approach can be modified to describe NQFTs. In Weinberg's approach, expectation values of time-ordered products of an interaction Hamiltonian density are an essential ingredient of interacting RQFTs, and in order for such quantities to be restricted Lorentz invariant, the interaction Hamiltonian density must commute at spacelike separated distances. Since time-ordering is always invariant under the symmetry group of a classical spacetime (due to the presence of an absolute temporal metric), this last requirement is unnecessary in NQFTs formulated in Weinberg's approach. I will argue that this explains why the spin–statistics and CPT theorems fail in NQFTs in Weinberg's approach.

(c) In Section 3.2.3, I argue that the best way to understand why the CPT and spin–statistics theorems fail for Lagrangian NQFTs is by considering the structural differences between relativistic and non-relativistic field equations. The former take the form of hyperbolic partial differential equations, whereas the latter take the form of parabolic partial differential equations.

This difference translates into differences in the types of properties that solutions to these equations possess, and this has a direct impact on the derivation of CPT invariance and the SSC.

(d) In Section 3.2.4, I point out that in the algebraic approach, an essential aspect of the proofs of the CPT and spin–statistics theorems is the property of vacuum separability, and that this property fails in the NQFT context.

The last section of Chapter 3 (Section 3.3) is devoted to an analysis of the intertheoretic relations between RQFTs on the one hand, and NQFTs and NQM on the other. This analysis begins with that described in Bain (2010) and supplements it with a discussion of the relation between the Poincaré and Galilei groups given in terms of Inonu and Wigner's (1953) concept of contraction. I will suggest that this provides the basis for an intertheoretic relation that maps the *kinematical* structure of RQFTs into the kinematical *and* (aspects of) the dynamical structure of NQFTs.

3.1 RQFTs versus NQFTs

At the most general level, the distinction between a relativistic quantum field theory (RQFT) and a non-relativistic quantum field theory (NQFT) is perhaps best characterized in terms of spacetime structure. By an RQFT I will mean a quantum field theory invariant under the actions of the symmetry group of a *Lorentzian* spacetime. By an NQFT, I will mean a quantum field theory invariant under the actions of the symmetry group of a *classical* spacetime. What follows are brief descriptions of these types of spacetimes.

A Lorentzian spacetime can be represented by a pair (M, g_{ab}), where M is a smooth four-dimensional (4-dim) differentiable manifold and g_{ab} is a pseudo-Riemannian metric with Lorentz signature (1, 3). This metric satisfies a compatibility condition $\nabla_a g_{ab} = 0$ and this condition determines a unique derivative operator and hence a unique curvature tensor R^a_{bcd}. Perhaps the best known example of a Lorentzian spacetime is Minkowski spacetime, characterized by a vanishing curvature tensor, $R^a_{bcd} = 0$, encoding spatiotemporal flatness. The symmetry group of Minkowski spacetime is the Poincaré group, which is generated by vector fields that Lie annihilate the Minkowski metric η_{ab}. Symbolically, we require $\pounds_x \eta^{ab} = 0$, where $\pounds_x$ is the Lie derivative associated with x^a. Intuitively, this means that the transformations between reference frames defined by the integral curves of the vector field x^a preserve the structure of the Minkowski metric. This structure famously entails that there is no unique way to separate time from space in Minkowski spacetime: any two observers moving inertially with respect to each other will disagree on the time interval between any two events, and on the spatial interval between any two events. In coordinate form, elements of the Poincaré group may be represented by transformations

$$x^\mu \mapsto x^{\mu\prime} = \Lambda^\mu_\nu x^\nu + d^\mu, \qquad \mu, \nu = 0, 1, 2, 3 \qquad (Poincaré)$$

where $\Lambda^{\mu}_{\nu} \in SL(2, \mathbb{C})$ is a pure Lorentz boost (encoded in a 2×2 complex matrix with unit determinate), and $d^{\mu} \in \mathbb{R}^4$ is a spacetime translation. The number of parameters needed to specify a Poincaré transformation is ten (six for a pure Lorentz boost, four for a spacetime translation).

Minkowski spacetime is not the only example of a Lorentzian spacetime. Another example is Vacuum Einstein spacetime, characterized by a vanishing Ricci tensor, $R_{ab} = 0$. (This entails spatiotemporal flatness only in conformally flat Lorentzian spacetimes, i.e., those in which the Weyl curvature tensor vanishes.) In general, Lorentzian spacetimes can possess different metrics and hence different curvature; but all are characterized by the same Lorentzian signature. This signature encodes the fundamental characteristic associated with relativity of "mixing" one temporal and three spatial degrees of freedom. More precisely, it entails that, while Lorentzian spacetimes may differ over large-scale curvature, they all "look like" Minkowski spacetime in the small. (In particular, the tangent space at every point in a Lorentzian spacetime is isomorphic to Minkowski vector space.) So we might say that, locally, a Lorentzian spacetime has the structure of Minkowski spacetime. Corresponding to different types of Lorentzian spacetimes, there can be different types of RQFTs, some in flat (Minkowski) spacetime, and others in curved Lorentzian spacetimes.

A classical spacetime is a spacetime that minimally admits absolute spatial and temporal metrics. More precisely, a classical spacetime may be represented by a tuple $(M, h^{ab}, t_a, \nabla_a)$, where M is a differentiable manifold, h^{ab} is a $(0, 1, 1, 1)$ symmetric tensor field on M identified as a spatial metric, t_a is a covariant vector field on M which induces a $(1, 0, 0, 0)$ temporal metric $t_{ab} = t_a t_b$, and ∇_a is a derivative operator associated with a (non-unique) connection on M and compatible with the metrics in the sense $\nabla_c h^{ab} = \nabla_a t_b = 0$. The spatial and temporal metrics are also required to be orthogonal in the sense $h^{ab} t_b = 0$. These conditions allow M to be decomposed into instantaneous 3-dim spacelike hypersurfaces parameterized by a global time function. In particular, they entail that the time interval between any two events is invariant, as well as the spatial distance between simultaneous events:

$$\begin{aligned} t_2 - t_1 &= const., \\ |\mathbf{x}_2 - \mathbf{x}_1| &= const., \quad \text{if } t_2 = t_1. \end{aligned} \tag{3.1}$$

As Lévy-Leblond (1971: 225) indicates, the most general *linear* transformations that preserve equation (3.1) form the Galilei group. But if linearity is dropped, larger symmetry groups are allowed. The most general classical spacetime symmetry group is generated by vector fields x^a that Lie annihilate h^{ab} and t_a. Symbolically, we require $\pounds_x h^{ab} = \pounds_x t_a = 0$, and again, this means that the transformations between reference frames defined by the integral curves of the vector fields x^a preserve the structure of the absolute spatial and temporal metrics. This entails that in any classical spacetime, there is always a unique way to separate time from space: any two observers moving inertially with respect to each other will always agree on the time interval between any two events, and on the spatial interval between any two simultaneous events. In this sense, space and time are *absolute* in a classical spacetime.

On the other hand, the compatibility conditions in a classical spacetime do not determine a unique curvature tensor. Additional constraints on the curvature may be imposed, and such constraints define different types of classical spacetimes. Two examples include neo-Newtonian spacetime, characterized by $R^a_{bcd} = 0$, encoding spatiotemporal flatness, and Maxwellian spacetime, characterized by $R^{ab}_{cd} = 0$, encoding a rotation standard (Bain, 2004: 348–352). The symmetries of neo-Newtonian spacetime form the Galilei group G generated by vector fields x^a that Lie annihilate the spatial and temporal metrics, and the connection. Symbolically, $\pounds_x h^{ab} = \pounds_x t_a = \pounds_x \Gamma^a_{bc} = 0$ (where Γ^a_{bc} is the connection defined by ∇_a), and in coordinate form,

$$\begin{aligned} \mathbf{x} &\mapsto \mathbf{x}' = R\mathbf{x} + \mathbf{v}t + \mathbf{a} \\ t &\mapsto t' = t + b \end{aligned} \qquad (Galilei)$$

where $R \in SO(3, \mathbb{R})$ is a constant orthogonal rotation matrix, $\mathbf{v}$, $\mathbf{a} \in \mathbb{R}^3$ are velocity boost and spatial translation vectors, and $b \in \mathbb{R}^1$ is a time translation. One requires 10 parameters to specify a Galilei transformation: three each for velocity boosts and spatial translations, three to specify a rotation matrix, and one for a time translation.

The symmetries of Maxwellian spacetime are given by the Maxwell group generated by vector fields x^a that Lie annihilate the spatial and temporal metrics and the rotational part of the connection. Symbolically, $\pounds_x h^{ab} = \pounds_x t_a = \pounds_x \Gamma^{ab}_c = 0$ (where $\Gamma^{ab}_c = h^{bd}\Gamma^a_{bc}$). In coordinate form,

$$\begin{aligned} \mathbf{x} &\mapsto \mathbf{x}' = R\mathbf{x} + \mathbf{c}(t) \\ t &\mapsto t' = t + b \end{aligned} \qquad (Maxwell)$$

where $R \in SO(3, \mathbb{R})$ is a constant orthogonal rotation matrix, $\mathbf{c}(t) \in \mathbb{R}^3$ is a time-dependent spatial boost vector, and $b \in \mathbb{R}^1$ is a time translation. One requires an infinite number of parameters to specify a Maxwell transformation: three to specify a rotation matrix, one for a time translation, and a continuum of parameters to specify the arbitrary Maxwell "boost" $\mathbf{c}(t)$.

A quick and dirty distinction between neo-Newtonian and Maxwellian spacetime can be given in terms of the way the absolute spatial slices are "rigged:" in neo-Newtonian spacetime, the rigging consists of "straight" trajectories, whereas in Maxwellian spacetime, it consists of "straight" and "curved" trajectories. More precisely, a neo-Newtonian connection can distinguish between a straight and a curved trajectory, whereas a Maxwellian connection cannot. Both connections can, however, distinguish between straight and curved trajectories, on the one hand, and "corkscrew" trajectories, on the other, i.e., in both spacetimes there is an absolute standard of rotation. Thus different classical spacetimes have the same metrical structure, but different curvature; and all are characterized by absolute spatial and temporal metrics. Corresponding to different types of classical spacetimes, there can be different types of NQFTs, some in flat (neo-Newtonian) spacetime, and others in curved classical spacetimes.

At this point, it should perhaps be stressed that the physically relevant RQFTs of interest are Poincaré-invariant QFTs (i.e., RQFTs in Minkowski spacetime); these are the type of RQFTs that figure into the standard model of particle physics. The physically relevant NQFTs of interest are Galilei-invariant QFTs (GQFTs). These are the type of NQFTs that figure into descriptions of many condensed matter systems. On the other hand, Christian (1997) has constructed a non-relativistic quantum field theory of gravity that is invariant under an extension of the Maxwell group (see, e.g., Bain, 2004). The theoretical significance of this NQFT will be considered in Section 3.3.

Irreducible Representations of the Poincaré and Galilei Groups

To further unpack the significance of invariance under a spacetime symmetry group, I'd like to focus on the examples of the Poincaré and Galilei groups. The action of the elements of a symmetry group $\mathcal{G}$ on a space of states is given in terms of a *representation* of $\mathcal{G}$ on the space. For our purposes, the latter is a vector space V, and a representation of $\mathcal{G}$ is a map U that takes elements of $\mathcal{G}$ to linear transformations on V in such a way that preserves the group product. Elements of V are referred to as *carriers* of the representation of $\mathcal{G}$. An *irreducible representation* (irrep) of $\mathcal{G}$ on V is one for which there is no subspace of V invariant under the action of the image of U, other than the zero subspace or V itself. Intuitively, an irrep cannot be divided into parts. One can show that an irrep is labeled uniquely by the eigenvalues of the Casimir invariants of $\mathcal{G}$'s Lie algebra, i.e., those elements of the Lie algebra that commute with all other elements.

Recall from Section 1.1.1 that, with the exception of the algebraic approach, the approaches to the CPT and spin–statistics theorems reviewed in Chapter 1 all represent the states of a quantum field theory by elements of a Hilbert space $\mathcal{H}$ up to phase (i.e., by unit rays). One then requires that such states be invariant under a projective representation of a spacetime symmetry group, i.e., a representation that is unique up to a phase. In the relativistic case (see, e.g., Weinberg, 1995: Chapter 2) one constructs projective irreps of the restricted Poincaré group $P_+^\uparrow$. These correspond to non-projective irreps of the universal covering $\tilde{P}_+^\uparrow$ of the restricted Poincaré group (obtained by replacing the restricted Lorentz subgroup $L_+^\uparrow$ of $P_+^\uparrow$ with its universal covering group $SL(2, \mathbb{C})$). Such irreps are uniquely labeled by their mass and spin, these being the eigenvalues of the Casimir invariants $P_\mu P^\mu$ and $S_\mu S^\mu$, $S_\mu = -1/2\, \varepsilon_{\mu\nu\rho\sigma} \mathcal{J}^{\nu\rho} P^\sigma$, of the Poincaré Lie algebra (generated by infinitesimal spacetime-translations P^μ and Lorentz boosts $\mathcal{J}^{\mu\nu}$). P_μ and S_μ admit representations in $\mathcal{H}$ as the 4-momentum and spin operators.

In the Galilei-invariant case, projective irreps of the Galilei group G correspond to non-projective irreps of the central extension of the universal covering of the Galilei group (Lévy-Leblond, 1971: 252). The universal covering $\tilde{G}$ of the Galilei group is obtained by replacing the rotation subgroup $SO(3, \mathbb{R})$ of G with its universal covering group $SU(2, \mathbb{C})$. The central extension $\tilde{G}^*$ of $\tilde{G}$ is then obtained by including a generator M in the Lie algebra of G that commutes with all other generators. Irreps of $\tilde{G}^*$ are uniquely labeled by their mass, internal energy, and spin, these being the eigenvalues of the Casimir invariants M, $U = H - (1/2m)\mathbf{P}^2$, and $\mathbf{S}^2 = (\mathbf{J} - (1/m)\mathbf{K} \times \mathbf{P})^2$ of the Lie algebra of $\tilde{G}^*$ (generated by infinitesimal time-translations H, space-translations $\mathbf{P}$, rotations $\mathbf{J}$, Galilei boosts $\mathbf{K}$, and the one-parameter phase group M). M, U and $\mathbf{S}$ admit representations in $\mathcal{H}$ as the mass, internal energy, and spin operators (Lévy-Leblond, 1967: 161, 163).

Note that one major difference between the Poincaré and Galilei groups is that the latter admits non-trivial extensions, whereas the former does not. This means, roughly, that to obtain a non-projective representation of the Poincaré group, we only had to move to its universal double covering. To obtain a non-projective representation of the Galilei group, on the other hand, we had to first move to its universal double covering and then extend the group by including a mass operator. The eigenvalues of the latter (the masses m) provide an additional label for the irreps that corresponds to a charge. This imposes a superselection rule on GQFT states (similar to superselection rules imposed on RQFT states that possess charges) which prohibits superpositions of states with different masses (Lévy-Leblond, 1967: 160).

3.2 A Plethora of Approaches to NQFTs

As Chapter 1 has shown, there are (at least) four conceptually distinct approaches to formulating the spin–statistics and CPT theorems in RQFTs. Thus to explain why these theorems fail for NQFTs it behooves us to work through each of these approaches and indicate in each case, where the failure occurs. This will initially require formulating NQFTs in each of these approaches. The strategy for the most part will be to first formulate Galilei-invariant QFTs in each approach, and then consider the extent to which this treatment can be extended to NQFTs in general.

3.2.1 Axiomatic NQFTs

An axiomatic treatment of GQFTs was given in Lévy-Leblond (1967). Table 3.1 compares these Lévy-Leblond axioms with the Wightman axioms for RQFTs.[1]

In general, the areas where the axioms differ are exactly those areas that involve spacetime structure. The most obvious difference in this regard is axiom 2. As discussed in Section 2.1, one way to understand axiom 3, LC, is as encoding the requirement that fields associated with causally separated regions of spacetime should be independent of each other. The way "causal separation" gets

Table 3.1 *Axioms for QFTs.*

RQFT Wightman axioms	GQFT Lévy-Leblond axioms
W1. *Fields.* The fundamental dynamical variables of the theory are local field operators that act on a Hilbert space $\mathcal{H}$ of states.	L1. *Fields.* The fundamental dynamical variables of the theory are local field operators that act on a Hilbert space $\mathcal{H}$ of states.
W2. *Poincaré invariance.* $\mathcal{H}$ admits a unitary projective representation of the restricted Poincaré group, under which the fields transform appropriately.[2]	L2. *Galilei invariance.* $\mathcal{H}$ admits a unitary projective representation of the Galilei group, under which the fields transform appropriately.[2]
W3. *Local commutativity* (LC). The fields (anti-) commute at spacelike separations.	L3. *Non-relativistic local commutativity.* The fields (anti-) commute at equal times for non-zero spatial separation.
W4. *Vacuum state.* There is a vector $\vert 0\rangle$ in $\mathcal{H}$ satisfying the following conditions: (i) $\vert 0\rangle$ is Poincaré-invariant. (ii) $\vert 0\rangle$ is cyclic for $\mathcal{H}$. (iii) The spectrum of the 4-momentum operator on the complement of $\vert 0\rangle$ is confined to the forward light cone (*spectrum condition* (SC)).	L4. *Vacuum state.* There is a vector $\vert 0\rangle$ in $\mathcal{H}$ satisfying the following conditions: (i) $\vert 0\rangle$ is Galilei-invariant. (ii) $\vert 0\rangle$ is cyclic for $\mathcal{H}$ within a given mass sector. (iii) The spectrum of the internal energy operator on the complement of $\vert 0\rangle$ and within a given mass sector is bounded from below (*spectrum condition* (SC)).

fleshed out depends on the structure of the associated spacetime; so one would expect axiom 3 to be different in the relativistic and non-relativistic contexts. Note that *non-relativistic local commutativity* only requires the existence of absolute spatial and temporal metrics, so it will be common to all formulations of axiomatic NQFTs, and not just GQFTs in particular.

Axiom 4 describes the vacuum state: in the field-theoretic context, this is the state of zero energy. The last property of this state, the *spectrum condition* (SC), as encoded in axiom (4iii), involves a restriction on the energies that other states of the theory can possess. In particular, axiom (W4iii) guarantees positivity of energy in Lorentz frames, and its counterpart axiom (L4iii) encodes the fact that, in non-relativistic mechanics, the potential energy of a single-particle state is a matter of convention. Technically, this is encoded in the fact that irreducible representations of the extended Galilei group that differ on their internal energies are projectively equivalent (Lévy-Leblond, 1971: 277). Another feature of the extended Galilei group requires the restriction to mass sectors in axiom (L4). Thus in the GQFT context, the constraints imposed by axiom (L4) can be explained by appeal to the structure of the spacetime symmetry group. While one would not expect this type of explanation to be available in the context of other classical spacetime symmetry groups, nevertheless it might be claimed that the constraints mandated by axiom (L4) should be imposed as a condition of physicality on all NQFTs, in so far as all such theories view mass as an absolute quantity distinct from energy.[3]

Thus to move from GQFTs to NQFTs, arguably, requires minimal modification of the Lévy-Leblond axioms. One simply replaces axioms (L2) and (L4i) with invariance under the appropriate classical spacetime symmetry group.

Lévy-Leblond (1967: 165) explains the failure of the spin–statistics and CPT theorems in axiomatic GQFTs as due to the failure of relativistic LC (axiom W3 in Table 3.1). Lévy-Leblond begins by considering a free field that satisfies the Schrödinger equation. This can be expressed as

$$\Phi(\mathbf{x}, t) = (2\pi)^{-3/2} \int d^3 p e^{(-iEt + i\mathbf{p}\cdot\mathbf{x})} A(\mathbf{p}) \tag{3.2}$$

where $A(\mathbf{p})$ is the annihilation operator for a free particle state of mass m.[4] Recall that in the extended Galilei group, mass plays the role of a charge. This suggests that a particle of mass m can be associated with an antiparticle of mass $-m$. Thus, for a free particle state of mass m and corresponding creation/annihilation operators $A^\dagger(\mathbf{p})$, $A(\mathbf{p})$, there corresponds a free antiparticle state of mass $-m$, and corresponding creation/annihilation operators, $B^\dagger(\mathbf{p})$, $B(\mathbf{p})$. Lévy-Leblond (1967: 165) now considers a field operator formed as a *linear combination* of $A(\mathbf{p})$, $B^\dagger(\mathbf{p})$:

$$\Phi(\mathbf{x}, t) = (2\pi)^{-3/2} \int d^3 p \left[\xi e^{(-iEt + i\mathbf{p}\cdot\mathbf{x})} A(\mathbf{p}) + \eta e^{(iEt - i\mathbf{p}\cdot\mathbf{x})} B^\dagger(\mathbf{p})\right] \tag{3.3}$$

Possible motivations for this expansion will be considered below. One can show that this field satisfies the equal time (anti-) commutation relations

$$[\Phi(\mathbf{x}, t), \Phi^\dagger(\mathbf{y}, t)]_\pm = \left(|\xi|^2 \pm |\eta|^2\right) \delta^{(3)}(\mathbf{x} - \mathbf{y}) \tag{3.4}$$

This expression vanishes everywhere except for $\mathbf{x} = \mathbf{y}$, regardless of the values of ξ and η. This means that *non-relativistic local commutativity* (axiom L3 in Table 3.1) is satisfied regardless of whether we choose commutation or anti-commutation relations for Φ. Moreover, it is satisfied even when $\eta = 0$, i.e., in the absence of antiparticles. Lévy-Leblond takes these results as an explanation of the failure of the SSC and CPT invariance in GQFTs:

> This situation is to be contrasted with the relativistic case where the requirements of local commutativity on a free field constructed as in [(3.3)] impose both the existence of a TCP operation (corresponding to the equal contribution of particles and antiparticles, i.e., $|\xi| = |\eta|$), and the spin–statistics relation, as has been shown in a very illuminating way, for this free-field case, by Weinberg (1964). (Lévy-Leblond, 1967: 165)

Note that the existence of equal numbers of particles and antiparticles does not quite establish CPT invariance in Weinberg's approach. Rather, the existence of equal numbers of particles and antiparticles follows in Weinberg's approach from the assumption that conserved charges exist. What establishes CPT invariance is a demonstration that the interaction Hamiltonian density is invariant with respect to the composition of C, P, and T (Weinberg, 1995: 244–246). More importantly, it's not entirely clear how results from Weinberg's approach can be appropriated by the axiomatic approach, since, at least on the surface, these approaches make different assumptions. Recall from Section 1.2.2 that in Weinberg's approach, one builds local quantum fields as linear combinations of creation and annihilation operators that satisfy RLI and relativistic LC in order to secure both RLI and *cluster decomposition* (CD) of the S-matrix. In particular, a relativistic scalar field associated with a charge q is given initially by (Weinberg, 1995: 204),

$$\phi(x) = (2\pi)^{-3/2} \int d^3p(2p^0)^{-1/2} \left[\kappa e^{ip\cdot x} a(\mathbf{p}) + \lambda e^{-ip\cdot x} a^{c\dagger}(\mathbf{p})\right] \qquad (3.5)$$

where $a(\mathbf{p})$ and $a^{c\dagger}(\mathbf{p})$ annihilate a free particle of charge $+q$ and create a free particle of charge $-q$, respectively. The analog of equation (3.4) then is

$$[\phi(x), \phi^\dagger(y)]_\pm = \left(|\kappa|^2 \pm |\lambda|^2\right) \Delta_+(x-y) \text{ for } (x-y) \text{ spacelike}, \qquad (3.6)$$

where $\Delta_+(x) = (2\pi)^{-3} \int d^3p(2p^0)^{-1} e^{ip\cdot x}$. Relativistic LC is then satisfied (for a non-vanishing field) *if and only if* $|\kappa| = |\lambda|$ and the bottom sign "–" is chosen. Thus the particle states associated with $\phi(x)$ obey Bose–Einstein statistics. The derivation of the SSC for fields with arbitrary spin follows in a similar fashion (Weinberg, 1995: 238).

Thus, with respect to the SSC, according to Lévy-Leblond, *non-relativistic local commutativity* does not place as great a constraint on the coefficients ξ, η in the non-relativistic expansion (3.3), as relativistic LC does on the coefficients κ, λ in the relativistic expansion (3.5). In the latter case, it is this constraint that, in Weinberg's approach, entails the SSC. Notice that the expansions (3.3) and (3.5) play crucial roles in this explanation. In Weinberg's approach, such an expansion follows from the assumption that the S-matrix satisfies CD. The additional assumption that the S-matrix satisfies RLI then uniquely fixes the values of the coefficients κ, λ, by requiring that the fields satisfy RLI and relativistic LC. In the axiomatic approach, on the other hand, while RLI and LC are imposed from the outset as axioms, there is no additional axiom (an analog, in this case, of CD of the S-matrix) that motivates the initial expansion of the fields as sums of Fock space creation/annihilation operators. This is a reflection of the generality of the Wightman axiomatic approach, namely, the fact that a model of the axioms need not take the form of a Fock space representation of the canonical (anti-) commutation relations. Thus Lévy-Leblond's appeal to Weinberg's approach in explaining why the spin–statistics theorem fails in axiomatic GQFTs doesn't seem to give the generality of the axiomatic approach its due.

Another way to see this is to note that expansions of the form of equation (3.5) in the relativistic context are typically motivated by the fact that relativistic field equations like the Klein–Gordon equation take the form of hyperbolic partial differential equations that have both positive energy and negative energy solutions. The introduction of antiparticle states allows the latter to be interpreted as positive energy solutions that describe antiparticles; thus, in particular $a(\mathbf{p})$ and $a^{c\dagger}(\mathbf{p})$ both are associated with positive energy states. The introduction of antiparticle states playing this role is unnecessary in the non-relativistic context in which the field equations take the form of parabolic partial differential equations that only possess positive energy solutions. On the other hand, Weinberg's motivation for introducing antiparticle states is based solely on the assumption that conserved charges exist; hence in order for the interaction Hamiltonian density to commute with the associated charge operator, the former must be constructed out of fields associated with particles and their antiparticles (Chapter 1, footnote 23). Again, in Weinberg's approach, fields and field equations play secondary roles to particle states and the S-matrix. Assumedly, this motivation might be adopted to underwrite Lévy-Leblond's expansion (3.3) by viewing mass as a conserved charge (as suggested by the extended Galilei group). However, this motivation seems to downplay the significance of fields, and this does not seem appropriate in the axiomatic approach.

Instead of appealing to Weinberg's proofs to explain why the spin–statistics and CPT theorems fail in axiomatic GQFTs, we should attempt to identity the steps in the *axiomatic* proofs that fail in the GQFT context. For the axiomatic spin–statistics theorem, there are two such steps. The first involves the derivation of the PT condition of Section 1.2.1 for Wightman functions. Recall that this condition allows one to show that, if the fields obey the wrong spin–locality connection, then they annihilate the vacuum. The second relevant step involves the application of the separating corollary to show that, if the fields annihilate the vacuum, then they must vanish (this is explained in the Appendix). For the axiomatic CPT theorem, the relevant step is simply the derivation of the PT condition for Wightman functions. This condition, combined with *weak local commutativity* (WLC), entails that the fields are CPT invariant. One can now show the following:

(a) In GQFTs, the PT condition cannot be derived for the Galilei-invariant analogs of Wightman functions (call them "Lévy-Leblond functions") due to the nature of the Galilei group.

(b) In GQFTs, the separating corollary does not hold.

Claims (a) and (b) demonstrate explicitly why the axiomatic spin–statistics and CPT theorems fail for axiomatic GQFTs. Thus, to the extent that the axiomatic spin–statistics and CPT theorems explain why the SSC and CPT invariance are essential properties of axiomatic RQFTs, claims (a) and (b) explain why these properties are not essential in axiomatic GQFTs. Moreover, both claims can be extended to NQFTs in general. The rest of this section will attempt to substantiate these remarks.

(a) The Failure of PT Invariance for Vacuum Expectation Values of Non-Relativistic Fields

Recall from Section 1.2.1 that an essential step in the axiomatic derivations of the SSC and CPT invariance for RQFTs was the demonstration that Wightman functions are invariant under a PT transformation. This result followed, in part, from the assumption that the fields were invariant under the (real) restricted Lorentz group $L_+^\uparrow$, even though the latter does *not* contain the PT transformation. The key to demonstrating PT invariance of Wightman functions was the additional assumption that the fields possess positive energy, i.e., the SC, axiom (W4iii) in Table 3.1. The conjunction of RLI and SC entails that Wightman functions can be analytically extended to complex analytic functions that can be shown to be invariant under the *complex* proper Lorentz group $L_+(\mathbb{C})$, and this group *does* contain the PT transformation.

Does this result hold in GQFTs? The Lévy-Leblond axioms include a non-relativistic version of SC, namely, axiom (L4iii) in Table 3.1. This axiom requires that the energy be bounded from below within each mass superselection sector (Lévy-Leblond, 1967: 161). Requardt (1982) has shown that this analog suffices to demonstrate complex-analyticity for the corresponding vacuum expectation values of products of Galilei-invariant fields (what were called "Lévy-Leblond functions" in claim (a)). However, PT invariance does not follow for Lévy-Leblond functions due to the nature of the Galilei group. In particular, neither the real nor the complex Galilei group has a component connected to the identity that contains the PT transformation. This result also holds for the symmetry group of any classical spacetime in general.

To make good on this explanation, I will need to recall a few facts about the Lorentz and Galilei groups. I will first review the relation between the restricted Lorentz group $L_+^\uparrow$ and its covering group $SL(2, \mathbb{C})$, and the relation between the complex proper Lorentz group $L_+(\mathbb{C})$ and its covering group $SL(2, \mathbb{C}) \otimes SL(2, \mathbb{C})$. I will then review the similar relations between the real and complex groups of 3-dim rotations, $SO(3, \mathbb{R})$ and $SO(3, \mathbb{C})$, and their covering groups, $SU(2, \mathbb{C})$ and $SL(2, \mathbb{C})$, respectively. This review will explicitly demonstrate that $L_+(\mathbb{C})$ contains the PT transformation, whereas $L_+^\uparrow$, $SO(3,\mathbb{R})$, and $SO(3, \mathbb{C})$ do not. I will then indicate why the results about $SO(3, \mathbb{R})$ and $SO(3, \mathbb{C})$ carry over to the Galilei group. Finally, I will indicate why these results hold, in general, for any classical spacetime symmetry group.

The relation between the real restricted Lorentz group $L_+^\uparrow$, and its covering group $SL(2, \mathbb{C})$, the group of complex 2×2 matrices with unit determinant, can be obtained in terms of a correspondence between real 4-vectors x^μ and complex 2×2 Hermitian matrices X,

$$X = \begin{pmatrix} x^0 + x^3 & x^1 - ix^2 \\ x^1 + ix^2 & x^0 - x^3 \end{pmatrix} = x^\mu \sigma_\mu, \quad \mu = 0, 1, 2, 3 \tag{3.7}$$

where σ_μ are the Pauli matrices.[5] Under this correspondence, the determinant of X is the Lorentz length of x^μ, $\det X = \eta_{\mu\nu} x^\mu x^\nu$, where $\eta_{\mu\nu}$ is the Minkowski metric. Thus, under this correspondence, a Lorentz transformation (i.e., a transformation on a 4-vector $x^\mu \mapsto x^{\mu'} = \Lambda^\mu_\nu x^\nu$ that preserves its Lorentz length) corresponds to a transformation on a complex 2×2 Hermitian matrix that preserves its determinant. This latter transformation is given by

$$X \mapsto X' = AXA^\dagger \tag{3.8}$$

where $A \in SL(2, \mathbb{C})$. One can show that this correspondence defines a 2:1 homomorphism from $SL(2, \mathbb{C})$ to $L_+^\uparrow$.[6] Note that the PT transformation $X \mapsto X' = -X$ cannot be implemented by the transformation (3.8), i.e., there are no matrices $A \in SL(2, \mathbb{C})$ for which $-X = AXA^\dagger$. This is a restatement of the fact that the restricted Lorentz group $L_+^\uparrow$ does not contain the PT transformation.

While the PT transformation is not contained in $L_+^\uparrow$, it is contained in the component of the complex Lorentz group connected to the identity, i.e., the proper complex Lorentz group $L_+(\mathbb{C})$. To see this, note first that the complex Lorentz group $L(\mathbb{C})$ can be thought of as the group of transformations that preserve the Lorentz length of complex 4-vectors. To any given complex 4-vector, one can again associate a complex 2×2 matrix X, however it need not be Hermitian. One is then led to considering transformations on such matrices that preserve their determinant, and these take the general form

$$X \mapsto X' = AXB^\dagger \tag{3.9}$$

where $A, B \in SL(2, \mathbb{C})$. One can then construct a 2:1 homomorphism between the product group $SL(2,\mathbb{C}) \times SL(2,\mathbb{C})$, consisting of pairs of matrices $\{A,B\}$, and the proper complex Lorentz group $L_+(\mathbb{C})$ (i.e., the component of $L(\mathbb{C})$ connected to the identity). Because two distinct matrices $\{A, B\}$ implement a complex Lorentz transformation, one can implement a PT transformation $X \mapsto -X$. In particular, there are two pairs of matrices $\{A, B\} \in SL(2,\mathbb{C}) \times SL(2,\mathbb{C})$ for which $-X = AXB^\dagger$; namely, $\{1,-1\}$ and $\{-1, 1\}$. Note, too, as Greenberg (2006b: 1540) demonstrates, that both of these PT transformations can be continuously connected to the identity $\{1,1\}$. In the former case, it suffices to let $B(\phi) = \mathrm{diag}(e^{i\phi/2}, e^{-i\phi/2})$; thus $B(0) = 1$ and $B(2\pi) = -1$, and as ϕ ranges from 0 to 2π, we smoothly connect $\{1,1\}$ and $\{1,-1\}$ (note that this path avoids the zero matrix, which does not correspond to a Lorentz transformation).

Thus by moving to the complex version of the Lorentz group, we are able to include the PT transformation in the component of the group connected to the identity. This is not the case for the group of 3-dim rotations. Both the group of real 3-dim rotations, $SO(3, \mathbb{R})$, and the group of complex 3-dim rotations, $SO(3, \mathbb{C})$, do not possess components connected to the identity that also contain the PT transformation. This can be shown, again, by considering their respective covering groups. Thus, to every real 3-vector x_i one can associate a traceless complex 2×2 Hermitian matrix given by

$$X_0 = \begin{pmatrix} x^3 & x^1 - ix^2 \\ x^1 + ix^2 & -x^3 \end{pmatrix} = x_i\sigma_i, \; i = 1,2,3 \tag{3.10}$$

One then notes that the determinant of X_0 is the Euclidean length of x_i, and this establishes a 2:1 map between transformations that preserve the Euclidean length of x_i (i.e., real 3-dim rotations) and transformations that preserve the determinant of traceless complex 2×2 Hermitian matrices. The latter are given by

$$X_0 \mapsto X_0' = UX_0U^\dagger \tag{3.11}$$

where $U \in SU(2, \mathbb{C})$, i.e., the group of complex 2×2 unitary matrices with unit determinant. Again, one can show that there is a 2:1 homomorphism between $SU(2, \mathbb{C})$ and $SO(3, \mathbb{R})$, i.e., the former is the double covering group of the latter. Moreover, since there is no $U \in SU(2, \mathbb{C})$ for which $-X_0 = UX_0U^\dagger$, the PT transformation is not contained in $SO(3, \mathbb{R})$. One can now show that the double covering group of $SO(3, \mathbb{C})$ is $SL(2, \mathbb{C})$, i.e., complex 3-vectors correspond to traceless complex 2×2 matrices, and the transformations that preserve the determinant of the latter are elements of $SL(2, \mathbb{C})$. Thus, in the case of 3-dim rotations, the covering group of the complex group is not a product group, as it is in the case of Lorentz transformations. Since there is no $A \in SL(2, \mathbb{C})$ for which $-X_0 = AX_0A^\dagger$, where X_0 is a traceless complex 2×2 matrix, the PT transformation is not contained in $SO(3, \mathbb{C})$.

To carry these results over to the Galilei group, it suffices to note that the real Galilei group G is homomorphic to $SO(3, \mathbb{R})$.[7] Thus since there is no PT transformation in $SO(3, \mathbb{R})$, there is no PT transformation in G. Now, to the extent that the complex Galilei group $G(\mathbb{C})$ is similarly homomorphic to $SO(3, \mathbb{C})$, and the latter does not contain a PT transformation, neither does the former.

In any classical spacetime in which the instantaneous spaces are Euclidean, the rotation group $SO(3, \mathbb{R})$ will continue to play the role it plays in the Galilei group; thus the symmetry group of such a spacetime will not possess a connected component that contains the PT transformation, even in the complex version of the group. Conceivably, there can be classical spacetimes with non-Euclidean instantaneous spaces. In these, $SO(3)$ will not, in general, appear as a factor group. However, the distinguishing characteristic of a classical spacetime is the presence of separate spatial and temporal metrics. The temporal metric requires that the group of time translations (isomorphic to $\mathbb{R}^1$) be a factor group. This means that there is no T transformation in the connected component of a real classical spacetime symmetry group, and hence no PT transformation.[8]

(b) The Failure of the Separating Corollary in Axiomatic NQFTs

The *separating corollary* is essential to the axiomatic proof of the spin–statistics theorem. It shows that, under certain conditions met by axiomatic RQFTs, the vacuum state is separating, i.e., for any field operator ϕ, if ϕ annihilates the vacuum, then ϕ is identically zero. The axiomatic spin–statistics theorem then reduces to the demonstration that, if the wrong spin–statistics connection (NSSC) is imposed on a field, then it annihilates the vacuum (see the Appendix for details). This section will show that the *separating corollary* does not hold for axiomatic GQFTs, and indeed, for axiomatic NQFTs in general. Thus even if these theories could be associated with complex-analytic vacuum expectation values that satisfied a PT condition, the axiomatic Spin–Statistics theorem would still fail.

The *separating corollary* states the following (Streater and Wightman, 1964: 139): let $\mathfrak{R}(\mathcal{O})$ be a von Neumann algebra of field operators associated with a bounded region $\mathcal{O}$ of spacetime. Then the vacuum is separating if the following conditions hold:

(i) The vacuum is cyclic for $\mathfrak{R}(\mathcal{O})$ (*local cyclicity*).
(ii) The causal complement $\mathcal{O}'$ of $\mathcal{O}$ is non-empty.
(iii) Relativistic *local commutativity* (LC) holds.

In condition (ii), the causal complement $\mathcal{O}'$ of a region $\mathcal{O}$ is the set of all points causally separated from points in $\mathcal{O}$. Conditions (i), (ii), and (iii) are satisfied by axiomatic RQFTs. Condition (i) is secured by the Reeh–Schlieder theorem, which demonstrates how local cyclicity follows from the SC (Wightman axiom (W4iii) in Table 3.1) (Streater and Wightman, 1964: 138). Condition (ii) is secured by the structure of Minkowski spacetime, provided one defines *causal separation* in Minkowski spacetime to be *spacelike separation.* Finally, condition (iii) is Wightman axiom (W3) in Table 3.1.

Does the *separating corollary* hold for axiomatic NQFTs? First note that condition (i) continues to hold for GQFTs. Requardt (1982) showed that local cyclicity follows from the non-relativistic version of the SC (Lévy-Leblond axiom (L4iii) in Table 3.1). Condition (iii), of course, fails by definition; however, just as there is a non-relativistic version of the SC, there is a non-relativistic version of LC; namely Lévy-Leblond axiom (L3) in Table 3.1. However, now we are faced with a choice: *non-relativistic local commutativity* only applies to *spatial* regions of classical spacetimes, i.e., regions with zero temporal extent. Thus, in an axiomatic NQFT, we need be a bit more explicit about the type of region we associate with a local algebra of field operators. In particular, one can consider two options:

(1) Associate local algebras with *spatiotemporal* regions, i.e., regions with non-zero spatial and temporal extent.
(2) Associate local algebras with *spatial* regions, i.e., regions with zero temporal extent.

In the relativistic context, this is a distinction that makes no difference. One can show that the local algebra associated with a spatial region $\mathcal{O}$ in Minkowski spacetime is the same as the local algebra associated with the domain of dependence $D(\mathcal{O})$ of $\mathcal{O}$ (Horuzhy, 1990: 40–41, Theorem 1.3.14).[9] The latter, which consists of points p for which any (inextendible) causal worldline through p intersects $\mathcal{O}$, typically is a spatiotemporal region. On the other hand, in the non-relativistic context, the distinction between (1) and (2) is non-trivial. However, in both cases, one can show that the *separating corollary* cannot be derived.

To see this, first suppose we adopt (1) for axiomatic NQFTs. Then Requardt's (1982) proof of local cyclicity goes through.[10] However, under reasonable assumptions, the causal complement of a spatiotemporal region of a classical spacetime is typically the empty set (barring topological mutants). In other words, all points outside a spatiotemporal region of a classical spacetime are typically causally connectible to some subset of points within that region. This assumes that the causal complement of a spatiotemporal region $\mathcal{O}$ of a classical spacetime consists of all points with zero temporal separation and non-zero spatial separation from all points in $\mathcal{O}$, which in turn assumes that infinite causal propagations are prohibited, but allows that finite causal propagations have no upper bound. These assumptions seem reasonable in so far as they follow from the requirement that simultaneous measurements be causally independent, which, arguably, motivates LC in the non-relativistic context. The upshot is that condition (ii) for the *separating corollary* is not met, and the vacuum is not separating.

Now suppose we adopt option (2) and associate local algebras of operators with *spatial* regions of classical spacetimes. Then Requardt's (1982) proof of local cyclicity fails, and condition (i) of the *separating corollary* is not met.

Thus, the non-relativistic vacuum is not cyclic for *spatial* local algebras, and hence not separating; and while it is cyclic for *spatiotemporal* local algebras, it is not separating. One can argue that both of these results are due to the spatiotemporal structure of classical spacetimes; in particular, to the existence of an absolute temporal metric. First, it is the simultaneity structure associated with an absolute temporal metric that trivializes the causal complement of a spatiotemporal region of a classical spacetime; and this, in turn, prevents the non-relativistic vacuum from being separating for local algebras associated with such regions. Second, in brief, proofs of the local cyclicity of the relativistic vacuum for spatial local algebras are based on the anti-local property, for spatial regions, of differential operators associated with relativistic field equations (Halvorson, 2001: 118–119, reviews the Klein–Gordon case). This anti-local property for spatial regions is in part a consequence of the fact that relativistic field equations are *hyperbolic*, reflecting the Lorentzian metrics of relativistic spacetimes. In contrast, the differential operators associated with non-relativistic field equations are *parabolic*, reflecting the degenerate metrics in classical spacetimes; in particular, these spacetimes contain separate temporal metrics.[11] It is this feature that prevents parabolic differential operators from being anti-local for spatial regions, and thus is the local cyclicity of the non-relativistic vacuum for spatial local algebras blocked.

3.2.2 NQFTs in Weinberg's Approach

Lévy-Leblond's explanation of why the spin–statistics theorem fails for axiomatic GQFTs appealed to Weinberg's demonstration of the SSC for RQFTs. In particular, Lévy-Leblond showed that imposing *non-relativistic local commutativity* on a linear combination of Galilei-invariant fields does not constrain them to possess the SSC, whereas Weinberg demonstrated that imposing relativistic LC on a linear combination of restricted Lorentz invariant fields does constrain them to possess the SSC. In Section 3.2.1, I argued that Lévy-Leblond's explanation is inadequate to the extent that an explanation of why the spin–statistics theorem fails for axiomatic GQFTs should identify the steps in the *axiomatic* proof of the theorem which fail for axiomatic GQFTs, and Lévy-Leblond's appeal to *Weinberg's* proof of the theorem fails to do this. On the other hand, one might be tempted to adopt Lévy-Leblond's demonstration as an explanation of why the spin–statistics theorem fails for GQFTs in Weinberg's approach. In particular, perhaps we can take Lévy-Leblond as demonstrating that it is the failure of relativistic LC by GQFTs that explains why the spin–statistics theorem fails for GQFTs in *Weinberg's* approach. However, I will argue that such an explanation is still inadequate insofar as it still conflates the axiomatic approach and Weinberg's approach. An appropriate explanation of why the spin–statistics theorem fails for GQFTs in Weinberg's approach should first formulate a GQFT in Weinberg's approach and then identify the steps in Weinberg's proof of the spin–statistics theorem that fail for such a GQFT. Lévy-Leblond's demonstration fails to do the former. I will argue that it is not the failure of relativistic LC that explains why Weinbergian GQFTs do not possess the SSC (this perhaps should be obvious to the extent that Weinberg views LC of fields not as an essential characteristic of a QFT, but as derivative of more fundamental assumptions); rather, the reason why Weinbergian GQFTs do not possess the SSC is due to the fact that time-ordering is always invariant under a classical spacetime symmetry group; thus imposing the latter on a time-ordered product does not place any constraints on the terms in the product in the way that imposing RLI on a time-ordered product does.

Recall from Section 1.2.2 that in Weinberg's approach, the basic object of an RQFT is an S-matrix which satisfies the following three conditions:

(1) The S-matrix is expressed as a perturbative power series expansion in time-ordered products of an interaction Hamiltonian density $\mathfrak{H}_{int}(x)$.

(2) The S-matrix is invariant under the restricted Lorentz group (RLI).

(3) The S-matrix satisfies *cluster decomposition* (CD).

These conditions apply with minimal modification to NQFTs. Condition (1) carries over without modification. Condition (2) can be replaced with invariance under a classical spacetime symmetry group. With respect to condition (3), recall that CD requires that the S-matrix factorizes $S_{\beta_i+\beta_j,\alpha_i+\alpha_j} = S_{\beta_i\alpha_i}S_{\beta_j\alpha_j}$ for clusters $|\alpha_i\rangle \rightarrow |\beta_i\rangle$ and $|\alpha_j\rangle \rightarrow |\beta_j\rangle$ of scattering events at a great spatial separations.

As Section 2.1.3 explained, this is a locality constraint in the sense of translating an independence condition that holds for clusters of scattering events, into the probabilities associated with these events. The independence condition on clusters can be phrased in either relativistic or non-relativistic terms. In the relativistic context, clustering occurs for spacelike separated scattering events, while in the non-relativistic context, clustering occurs for scattering events in the limit as their equal-time spatial separation goes to infinity.

Thus Weinberg's approach applied to the non-relativistic context identifies the basic object of an NQFT as an S-matrix which satisfies the following conditions:

(1′) The S-matrix is expressed as a perturbative power series expansion in time-ordered products of an interaction Hamiltonian density $\mathfrak{H}_{int}(x)$.

(2′) The S-matrix is invariant under the symmetry group of a classical spacetime.

(3′) The S-matrix satisfies *cluster decomposition* (CD).

Recall that, for RQFTs, conditions (1) and (2) require the interaction Hamiltonian density to be a Lorentz scalar and to commute at spacelike separated distances:

$$[\mathfrak{H}_{int}(x), \mathfrak{H}_{int}(y)] = 0, \text{ for spacelike } (x-y). \tag{3.12}$$

Condition (3.12) is required to guarantee that the time-ordered products of $\mathfrak{H}_{int}(x)$ that appear in the expression for the S-matrix are restricted Lorentz invariant. Weinberg then demonstrates that "the only known way" (1964: 1318) to make these constraints imposed by conditions (1) and (2) compatible with condition (3) is to construct $\mathfrak{H}_{int}(x)$ out of particular linear combinations of creation and annihilation operators that act on a Fock space comprised of particle states given by finite irreducible representations of the restricted Lorentz group. These linear combinations take the form of restricted Lorentz invariant fields that, due to condition (3.12), must satisfy relativistic LC, and the latter guarantees they possess the SSC.

Now let's consider the extent to which this reasoning applies to the NQFT case. Conditions (1′) and (2′) can be made compatible by requiring the interaction Hamiltonian density $\mathfrak{H}_{int}(x)$ to be an invariant scalar of the appropriate classical spacetime symmetry group. However, condition (3.12) is no longer needed. In general, time-ordering is always invariant under the isometry group of a classical spacetime, due to the presence of an absolute temporal metric. Weinberg himself stresses this difference: "This condition [viz., (3.12)] has no counterpart for non-relativistic systems, for which time-ordering is always Galilean-invariant. *It is this condition that makes the combination of Lorentz invariance and quantum mechanics so restrictive.*" (Weinberg, 1995: 145). Note, again, that this difference holds not just for Galilean-invariant QFTs, but for any QFT invariant under a symmetry group that leaves invariant an absolute temporal metric.

Condition (3′) can be made compatible with conditions (1′) and (2′) by, again, constructing $\mathfrak{H}_{int}(x)$ out of linear combinations of creation and annihilation operators that now act on a Fock space of particle states given by finite irreducible representations of a classical spacetime symmetry group. These linear combinations can be identified as fields, but we no longer must insist that they satisfy (in this case) *non-relativistic local commutativity*. Thus there are no constraints imposed by conditions (1′), (2′), and (3′) that entail the fields we identify with our NQFT must possess the SSC. We reach the same conclusion that Lévy-Leblond reaches, but not by imposing LC on our fields; rather, our conclusion is entailed by the imposition of a classical spacetime symmetry group on the *S*-matrix. In particular, models of Weinbergian NQFTs that do not possess the SSC are possible due the fact that Weinbergian NQFTs need not satisfy condition (3.12).

3.2.3 Lagrangian NQFTs

What explains the failure of the spin–statistics and CPT theorems in Lagrangian NQFTs? Let's first recall some facts about the Lagrangian approach to RQFTs (Section 1.2.3). This approach begins with a Lagrangian density for a classical field theory. The fields are obtained as solutions to the corresponding Euler–Lagrange equations of motion. As Section 3.2.1 observed, for a restricted Lorentz invariant Lagrangian density, the field equations take the form of hyperbolic partial differential equations, whereas for a Lagrangian density invariant under the symmetry group of a classical spacetime, the field equations take the form of parabolic partial differential equations. This difference is the key to understanding why the CPT and spin–statistics theorems fail in Lagrangian NQFTs, as I shall now argue.

Briefly, hyperbolic partial differential equations admit both positive and negative energy solutions, whereas parabolic partial differential equations only admit positive energy solutions. This difference entails, among other things, that a Hamiltonian operator built out of quantum fields obtained as second-quantized solutions to a hyperbolic partial differential equation can possess negative energy eigenvalues, whereas a Hamiltonian operator built out of quantum fields obtained as second-quantized solutions to a parabolic partial differential equation will possess only positive energy solutions.

To see how this difference explains why CPT invariance and the SSC are not essential properties in NQFTs, let's first consider the Lagrangian approach to the spin–statistics theorem in RQFTs. This approach takes the classical field solutions and second-quantizes them. This involves a procedure in which each field is expanded in a Fourier series with coefficients determined by the particular representation of the restricted Lorentz group that the field transforms under; the coefficients hence encode the spin of the field. One then identifies these coefficients as creation and annihilation operators that act on a Fock space of multi-particle states. One is then faced with a choice of statistics to impose on the field: either commutation relations or anti-commutation relations can be imposed on the creation/annihilation operators at spacelike separated distances (i.e., one imposes condition (1.1) of Section 1.1.2). For an integer spin field (like a spin-0 Klein–Gordon field), the following results can then be demonstrated (Kaku, 1993: 90; Peskin and Schroeder, 1995: 13–14):

(i) The Hamiltonian possesses only positive eigenvalues.
(ii) The commutator of the field and its conjugate vanishes at spacelike separated distances.
(iii) The anticommutator of the field and its conjugate does not vanish.

Result (i) requires the introduction of antiparticle states in order to reinterpret negative energy solutions (allowed by hyperbolic field equations) as positive energy antimatter solutions. Result (ii) is interpreted as encoding *causality*, i.e., at spacelike separated distances, observables must commute to assure causal independence of measurements. To identify result (ii) as enforcing *causality* is to identify integer-spin fields as observables. Thus for an integer-spin field, imposing the NSSC (i.e., imposing anticommutation relations) entails a violation of *causality*. This is entailment (b) of Section 1.2.3.

Now suppose one starts with a restricted Lorentz invariant Lagrangian density that contains a half-integer-spin field. When we second-quantize this field, the following results can be demonstrated (Kaku, 1993: 86; Peskin and Schroeder, 1995: 52):

(i′) The Hamiltonian can possess negative eigenvalues.
(ii′) The commutator of the field and its conjugate does not vanish.
(iii′) The anticommutator of the field and its conjugate vanishes at spacelike separated distances.

In this case, the introduction of antiparticle states alone is not sufficient to guarantee that the field possesses positive energy. If one chooses to encode statistics with anticommutators, then *causality* can be upheld by allowing bilinears in half-integer-spin fields to count as observables. The choice of imposing anticommutators on the field also addresses result (i′) by allowing the negative energy term to be reinterpreted as a positive energy term (under normal ordering). Thus, for a half-integer-spin relativistic field, imposing the NSSC (i.e., imposing commutation relations) entails either a violation of *causality* or a violation of the SC. This is entailment (c) of Section 1.2.3.

Now consider how these results fair in the non-relativistic context. One begins with a non-relativistic classical Lagrangian density, i.e., a Lagrangian density invariant under the symmetry group of a classical spacetime, and one attempts to second-quantize its fields. Again, this involves first expanding each field as a Fourier series with expansion coefficients determined by the representation of the classical spacetime symmetry group that the field transforms under, and then identifying these coefficients with creation/annihilation operators that act on a non-relativistic multi-particle Fock space. For both integer-spin and half-integer-spin fields, one can then demonstrate the following results:

(i*) The Hamiltonian possesses only positive eigenvalues.
(ii*) The commutator of the field and its conjugate vanishes at equal times for large spatial distances.
(iii*) The anticommutator of the field and its conjugate vanishes at equal times for large spatial distances.

In this non-relativistic case, we can still identify a version of *causality*; namely,

The observable quantities associated with an NQFT commute at equal times for large spatial distances. (Non-relativistic causality)

Non-relativistic causality is enough to secure causal independence of simultaneous measurements in classical spacetimes: it assumes that infinite causal propagations are prohibited, but allows that finite causal propagations have no upper bound. However, neither *Non-relativistic Causality* nor the non-relativistic SC (axiom (L4iii) in Table 3.1) place the same type of constraints on the choice of statistics for non-relativistic fields, as their relativistic analogues do for relativistic fields.

Thus the fields that appear in a non-relativistic Lagrangian density need not possess the SSC. What about CPT invariance? Recall from Section 1.2.3 that, in the Lagrangian approach, CPT invariance is proven for a theory's Hamiltonian density under the assumptions that the SSC holds for the fields that appear in it, that these fields are restricted Lorentz invariant, and that they appear in a local Hermitian normal-ordered Langrangian density. Each of these assumptions fails in the non-relativistic context: the SSC need not hold, the fields are not restricted Lorentz invariant, and a non-relativistic Lagrangian density need not be Hermitian.[12] With respect to the latter, note that the Galilean-invariant quantum field (3.2) discussed in Section 3.2.1 is not Hermitian (Lévy-Leblond, 1967: 163): it transforms like a field with mass m under a Galilean transformation, whereas its adjoint transforms like a field with mass $-m$. Thus any Lagrangian density in which Galilean-invariant fields appear will not be Hermitian.

3.2.4 Algebraic NQFTs

Recall from Section 1.4 that the basic object in the algebraic approach to RQFTs is a net of von Neumann algebras, $\mathcal{O} \mapsto \mathfrak{R}(\mathcal{O})$, that assigns a local algebra of observables $\mathfrak{R}(\mathcal{O})$, represented by self-adjoint operators on a Hilbert space $\mathcal{H}$, to every double-cone region $\mathcal{O}$ of Minkowski spacetime. The local algebras are required to satisfy the Haag–Araki axioms reproduced in the left-hand column

Table 3.2 *Axioms for algebraic QFTs.*

RQFT Haag–Araki algebraic axioms	NQFT algebraic axioms
HA1. *Isotony.* $\mathcal{O}_1 \subset \mathcal{O}_2 \Rightarrow \mathfrak{R}(\mathcal{O}_1) \subset \mathfrak{R}(\mathcal{O}_2)$.	1. *Isotony.* $\mathcal{O}_1 \subset \mathcal{O}_2 \Rightarrow \mathfrak{R}(\mathcal{O}_1) \subset \mathfrak{R}(\mathcal{O}_2)$.
HA2. *Poincaré invariance.* To every restricted-Poincaré transformation g there corresponds an automorphism α_g such that $\alpha_g \mathfrak{R}(\mathcal{O}) = \mathfrak{R}(g\mathcal{O})$.	2. *Non-relativistic invariance.* To every $g \in \mathcal{G}$ where $\mathcal{G}$ is a classical spacetime symmetry group, there corresponds an automorphism α_g such that $\alpha_g \mathfrak{R}(\mathcal{O}) = \mathfrak{R}(g\mathcal{O})$.
HA3. *Algebraic causality.* For $A_1 \in \mathfrak{R}(\mathcal{O}_1)$ and $A_2 \in \mathfrak{R}(\mathcal{O}_2)$, and $\mathcal{O}_1$, $\mathcal{O}_2$ spacelike separated, $[A_1, A_2] = 0$.	3. *Non-relativistic algebraic causality.* For $A_1 \in \mathfrak{R}(\mathcal{O}_1)$ and $A_2 \in \mathfrak{R}(\mathcal{O}_2)$ and $\mathcal{O}_1$, $\mathcal{O}_2$ spatially separated at equal times, $[A_1, A_2] = 0$.
HA4. *Generating property.* The collection of all local algebras $\mathfrak{R}(\mathcal{O})$ over all spacetime regions $\mathcal{O}$ generates a von Neumann algebra $\mathfrak{R}$.	4. *Generating property.* The collection of all local algebras $\mathfrak{R}(\mathcal{O})$ over all spacetime regions $\mathcal{O}$ generates a von Neumann algebra $\mathfrak{R}$.

of Table 3.2. These axioms apply with minimal modification to non-relativistic NQFTs. Thus, let an algebraic NQFT be a net of von Neumann algebras, $\mathcal{O} \mapsto \mathfrak{R}(\mathcal{O})$, that assigns a local algebra $\mathfrak{R}(\mathcal{O})$ to every open region of a classical spacetime (as Section 3.2 indicated, such regions can either be spatial or spatiotemporal). Axioms HA1 and HA4 of Table 3.2 can be adopted with no modification. Axiom HA2, *restricted Poincaré invariance*, must, of course be replaced with invariance under a classical spacetime symmetry group. Axiom HA3, *algebraic causality*, must also be modified; one can simply replace "spacelike separation" with "spatial separation at equal times." The resulting set of algebraic axioms for NQFTs is reproduced in the right-hand column of Table 3.2.

As discussed in Sections 1.4 and 2.2.1, Guido and Longo (1995) were able to show that, in the relativistic case, HA1, HA3, *additivity, modular covariance* (MC), and *normal commutation relations* are enough to derive the SSC and CPT invariance for DHR representations of $\mathfrak{R}$. In particular, HA2 is not necessary. Briefly, *additivity* entails that the vacuum state Ω is cyclic (this is the content of the algebraic formulation of the Reeh–Schlieder theorem); cyclicity combined with *algebraic causality* entails the vacuum is separating (this is the content of the separating corollary). A local von Neumann algebra with a cyclic and separating vacuum vector Ω possesses modular operators and a modular conjugation operator that leave Ω and the algebra invariant[13] (this is the content of the Tomita–Takesaki theorem). MC then asserts that the modular operators of a local algebra associated with a wedge region of spacetime implement $(1 + 1)$-dim Lorentz boosts; and this allowed Guido and Longo to construct a unique positive energy representation of the (3 + 1)-dim Poincaré group with respect to which DHR representations are CPT invariant and possess the SSC (the latter, provided statistics is encoded via *normal commutation relations*). On the other hand, as observed in Section 2.4, it may be disingenuous to say that this procedure indicates that CPT invariance and the SSC are independent of RLI, since the latter is derived in the process of deriving the former. This makes the exercise of determining where this derivation breaks down in the appropriate NQFT context important.

In the non-relativistic context, we can assume *additivity* without any need for modification. In addition, one presumes that Requardt's (1982) non-relativistic version of the Reeh–Schlieder theorem holds (albeit in an algebraic formulation). This entails that an algebraic non-relativistic vacuum state is locally cyclic for $\mathfrak{R}$. But recall from Section 3.2.1 that the vacuum state of an NQFT is not in general separating. Thus Tomita–Takesaki modular theory cannot in general be applied to NQFTs, and this means in particular that there is no non-relativistic analog of MC.

To see this in a bit more detail, consider the role that vacuum separability plays in Tomita–Takesaki theory. One starts with a pair $(\mathfrak{R}, \Omega)$ consisting of a von Neumann algebra $\mathfrak{R}$ that acts on a Hilbert space $\mathcal{H}$ with a cyclic and separating vacuum vector Ω. Recall that a state Ω is *cyclic* for $\mathfrak{R}$ just when $\{A\Omega : A \in \mathfrak{R}\}$ is dense in $\mathcal{H}$ (i.e., acting on Ω with any element of $\mathfrak{R}$ approximates any element of $\mathcal{H}$); and a state Ω is *separating* for $\mathfrak{R}$ if $A \in \mathfrak{R}$ and $A\Omega = 0$ imply $A = 0$. This entails $A\Omega \neq 0$, and thus $\langle \Omega, A^*A\Omega \rangle > 0$, for all $A \neq 0$. Given such a pair $(\mathfrak{R}, \Omega)$, one can define an antilinear operator $S_0 \in \mathfrak{R}$ with the following properties

$$S_0 : \mathfrak{R}\Omega \to \mathfrak{R}\Omega, \quad A\Omega \mapsto S_0(A\Omega) := A^*\Omega \tag{3.13}$$

S_0 can be thought of as the operator on the subset $\mathfrak{R}\Omega$ of $\mathcal{H}$ which implements the $*$-operation on $\mathfrak{R}$. One can then express the closure $S = S_0^{**}$ of S_0 in terms of its polar decomposition $S = \mathcal{J}\Delta^{1/2} = \mathcal{J}\Delta^{1/2}\mathcal{J}$, where Δ is referred to as the modular operator and $\mathcal{J}$ the modular conjugation of $\mathfrak{R}$. Then the Tomita–Takesaki theorem states that $\Delta^{it}\mathfrak{R}\Delta^{-it} = \mathfrak{R}$, and $\mathcal{J}\mathfrak{R}\mathcal{J} = \mathfrak{R}'$, and we have the framework within which to define MC. According to Verch, cyclicity and separability of the vacuum Ω are required in order for S_0 to be well-defined:

> By cyclicity of Ω for $\mathfrak{R}$, the set $\mathfrak{R}\Omega = \{A\Omega : A \in \mathfrak{R}\}$ is a dense linear subset of $\mathcal{H}$, so the operator is densely defined; furthermore to assign the value $A^*\Omega$ to the vector $A\Omega$ in the domain of [S_0] is a well-defined procedure in view of the assumption that Ω is separating for $\mathfrak{R}$. (Verch, 2006: 150)

Thus, given an algebraic NQFT satisfying axioms 1 and 3 of Table 3.2 and *additivity*, one cannot define a non-relativistic correlate of MC. This is due to the failure of a non-relativistic vacuum state to be separating, which entails that one cannot define an operator S_0 on the subset $\mathfrak{R}\Omega$ of $\mathcal{H}$ which implements the $*$-operation on $\mathfrak{R}$. Thus one cannot define modular operators and a modular conjugation operator in general for an algebraic NQFT; hence, an algebraic proof of the spin–statistics and CPT theorems for NQFTs, along the lines of Guido and Longo (1995), cannot be constructed.

Note finally that the failure of a non-relativistic vacuum state to be separating also explains why algebraic proofs of the spin–statistics and CPT theorems based on the alternative assumptions of MPCT, CGMA, and CMG (as discussed in Section 2.2.1) fail in algebraic NQFTs, insofar as all of these proofs also rely on the existence of modular data $(\Delta(t), \mathcal{J})$ for a given pair $(\mathfrak{R}, \Omega)$.

3.3 Intertheoretic Relations

In the previous section, explanations were offered for why the spin–statistics and CPT theorems fail in NQFTs. These explanations were offered within the context of each of the four approaches to RQFTs reviewed in Section 1.2. This section seeks to provide a framework in terms of which the intertheoretic relations between RQFTs and NQFTs can be understood. The reason for this is twofold. First, an explanation of the SSC and CPT invariance may depend, perhaps in part, on the relations between RQFTs, in which these properties are essential, and NQFTs, in which they are not essential. This will be the subject of Chapter 5. Second, the framework that describes the relation between RQFTs and NQFTs can be extended to encompass relativistic and non-relativistic quantum mechanics (with finite degrees of freedom), and relativistic and non-relativistic classical (i.e., non-quantum mechanical) theories with infinite degrees of freedom. These theories are important as we'll see in Chapter 4 since there is a large body of literature that seeks to derive the spin–statistics and CPT theorems within them.

3.3.1 The Bronstein Hypercube

The framework for intertheoretic relations I'd like to consider is embodied in the Bronstein cube.[14] This is a diagrammatic representation of the relations between fundamental theories in physics. It takes the form of a cube with axes representing the Newtonian gravitational constant G, Planck's constant $\hbar$, and the inverse speed of light $1/c$ (see Figure 3.1).

The vertices of the cube are meant to represent the following theories:

(a) Classical mechanics (CM)
(b) Special relativity (SR)
(c) General relativity (GR)
(d) Newton–Cartan gravity (NCG)[15]
(e) Newtonian quantum gravity (NQG)[16]
(f) Galilei-invariant quantum mechanics (GQM)
(g) Relativistic quantum field theory (RQFT)
(h) Relativistic quantum gravity (QG)

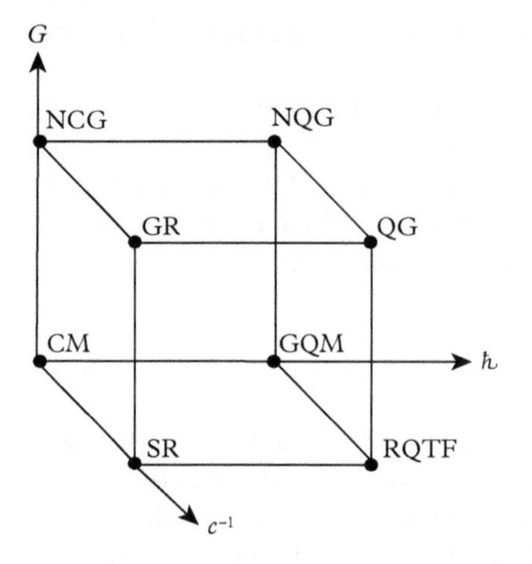

Figure 3.1 *The Bronstein cube. (From Bain, 2010: 141. Reprinted with permission from Springer Verlag.)*

Schematically, these theories can be described by their coordinates $(G, \hbar, 1/c)$. GR, for instance, may be given the coordinates (1, 0, 1), indicating that G and $1/c$ are "turned on," whereas $\hbar$ is "turned off." The cube thus entails that there are three distinct approaches to constructing QG: quantizing GR (epitomized in background independent approaches like loop quantum gravity); "turning on" gravity in an RQFT (epitomized in background dependent approaches like string theory); and the approach, novel to Christian (1997), of relativizing NQG.

Without further ado, we should be a bit wary about taking this representation of the relations among fundamental theories too literally. Under closer inspection, two types of problems arise. The first has to do with how the various limits relate to each other, and the second has to do with the definitions of the limits individually. First, as Stachel (2003: 21) notes, it is not clear if the limits commute with each other. For instance, turning on gravity in an RQFT may not lead to the same theory of quantum gravity as turning on $\hbar$ in general relativity. Moreover, one can let $G \to 0$, $h \to 0$ while keeping their ratio $\hbar/G$ (the Planck mass squared) constant; or one can let them go to zero at different rates, so that $\hbar/G$ goes to zero or infinity. These differences are non-trivial. For instance, the limit characterized by $G \to 0$, $h \to 0$, $\hbar/G$ = constant, is the "relative locality" limit of Amelino-Camelia et al. (2014). This limit describes theories characterized by a non-linear composition law for momenta (analogous to the non-linear composition law for velocities in special relativity), and thus should not be identified with the "SR" vertex in Figure 3.1.

The second type of problem concerns the nature of the individual limits:

(i) First, the $1/c \to 0$ limit that turns off relativity might initially be thought of as an Inonu–Wigner contraction of the Poincaré group to obtain the Galilei group (Inonu and Wigner, 1953). But this contraction actually consists of two limiting procedures: a low-velocity limit and a limit in which spacelike intervals are small compared to timelike intervals (Bacry and Lévy-Leblond, 1968 refer to the combination of these limits as a "speed–space" contraction). And even these two limits do not uniquely determine a non-relativistic counterpart of a given relativistic theory. For instance, there are two distinct non-relativistic limits, construed in this way, of the Maxwell equations (Holland and Brown, 2003). Moreover, the $1/c \to 0$ link between general relativity (GR) and Newton–Cartan gravity (NCG) cannot be described by a group contraction. On the one hand, the Poincaré group is not the symmetry group associated with GR (under one interpretation, the latter is the group Diff(M) of diffeomorphisms on the spacetime manifold M). On the other hand, there is more than one version of NCG, depending on how the geometrization procedure is carried out. One of these versions can indeed be shown to be a $1/c \to 0$ limit of GR, but this version does not have the Galilei group as its symmetry group (see, e.g., Bain, 2004: 365).

(ii) The $G \to 0$ limit might be associated simply with setting G to zero in the relevant dynamical equation (thus turning off gravity); but this would make the link between general relativity (GR) and special relativity (SR) problematic. Setting G to zero in the Einstein equations results in a Ricci-flat ($R_{ab} = 0$) Lorentzian spacetime, whereas Minkowski spacetime, the spacetime of SR, is spatiotemporally flat ($R^a_{bcd} = 0$). (One can show that Ricci-flatness only entails spatiotemporal flatness in conformally flat 4-dim spacetimes, for which the Weyl tensor vanishes.) This problematizes the other $G \to 0$ links as well, in so far as there can be Ricci-flat classical spacetimes other than neo-Newtonian spacetime, which, presumably, is the spacetime of classical mechanics (CM) and Galilei-invariant quantum mechanics (GQM).

(iii) Finally, one might describe the $\hbar \to 0$ limit as the inverse of quantization. But just how the quantization procedure should be characterized is far from settled. For instance, the quantization procedure that represents the link between special relativity (SR) and relativistic quantum field theory (RQFT) is not unique: for a theory of a classical relativistic field with infinite degrees of freedom, the failure of the Stone–von Neumann theorem entails that there are uncountably many unitarily inequivalent representations of the canonical (anti-) commutation relations of the corresponding QFT. Furthermore, inequivalent quantizations are not only associated with systems with infinite degrees of freedom; they also arise for finite systems with topologically non-trivial state spaces.[17] This problematizes the link between CM and GQM, as well as the link between NCG and Newtonian quantum gravity (NQG) (in the latter case, for topologically trivial gravitational fields, an appeal to the unique global time function in the appropriate classical spacetime serves to pick out a unique method of quantization).

In addition to these concerns, there is a deeper structural problem. Consider the link between NQG and GQM. NQG takes the form of Christian's (1997) NQFT of gravity in a curved classical spacetime. Turning off gravity should result in an NQFT in a Ricci-flat classical spacetime, and the scope of such a theory is larger than the scope of GQM, assuming the latter refers to GQM with finite degrees of freedom. Moreover, turning off "relativity" in an RQFT should result in an NQFT, and, again, the scope of "NQFT" in this context seems wider than the scope of "GQM." NQFTs are theories with *infinite* degrees of freedom, and they are invariant under the symmetries of a classical spacetime which *need not* be neo-Newtonian spacetime, the spacetime associated with the Galilei group.

These concerns stem from the fact that NQFTs are missing from the cube. NQFTs may be thought of as appropriately qualified $1/c \rightarrow 0$ limits of RQFTs. Suppose we relabel the GQM vertex as NQM and restrict its referent to non-relativistic quantum theories with finite degrees of freedom (i.e., finite quantum theories invariant under the symmetry group of a classical spacetime). Then, for N = degrees of freedom, NQMs may be thought of, schematically, as the "inverse thermodynamic" limit $N \rightarrow 0$ of NQFTs. This limit is intended to be applicable to quantum theories independently of classical theories and vice versa, i.e., it is intended to be "orthogonal" to the $h \rightarrow 0$ limit. Thus, for instance, it should also hold between a classical theory with an infinite number of degrees of freedom (a non-relativistic classical field theory, for instance), and a classical theory with finite degrees of freedom (a non-relativistic classical theory of a finite number of particles, for instance). What it informally suggests is that the Bronstein cube should be replaced by a 4-dim hypercube with an additional axis representing degrees of freedom N. Suppressing the G-dimension, we then have Figure 3.2.

The vertices in Figure 3.2 represent the following theories:

(a) Non-relativistic classical mechanics with finite degrees of freedom (NCM)
(b) Relativistic classical mechanics with finite degrees of freedom (RCM)
(c) Non-relativistic classical field theory (NCFT)

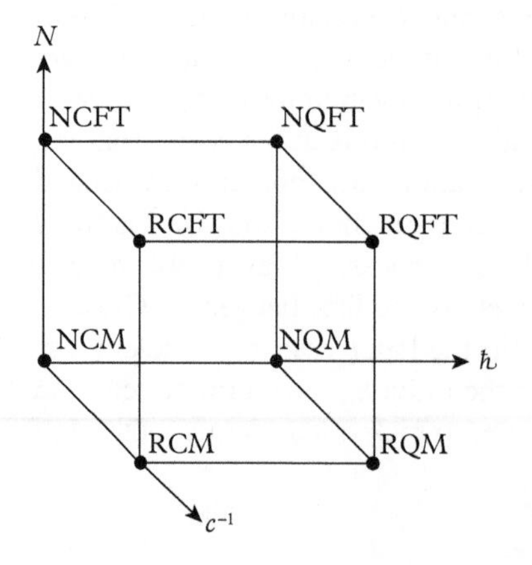

Figure 3.2 *Relations between theories in the absence of gravity. (From Bain, 2010: 144. Reprinted with permission from Springer Verlag.)*

(d) Relativistic classical field theory (RCFT)
(e) Non-relativistic quantum mechanics with finite degrees of freedom (NQM)
(f) Relativistic quantum mechanics with finite degrees of freedom (RQM)
(g) Non-relativistic quantum field theory (NQFT)
(h) Relativistic quantum field theory (RQFT)

The important distinction is between theories (classical and quantum, relativistic and non-relativistic) with infinite degrees of freedom, and theories (classical and quantum, relativistic and non-relativistic) with finite degrees of freedom. Theories in Bronstein hypercube space are coordinatized by 4-tuples $(G, h, 1/c, N)$. There are now four distinct approaches to constructing relativistic QG: quantizing the classical field theory of GR, with coordinates (1, 0, 1, 1); "turning on" gravity in an RQFT with coordinates (0, 1, 1, 1); "relativizing" a non-relativistic QFT of gravity (such as Christian's NQG) with coordinates (1, 1, 0, 1); or "taking the thermodynamic limit" of a relativistic quantum theory of gravity with a finite number of degrees of freedom, with coordinates (1, 1, 1, 0).

3.3.2 The Speed–Space Contraction of the Poincaré Group

With an eye toward the relation between RQFTs and NQFTs, I'd now like to consider the $1/c \to 0$ limit in a bit more detail. Both the Poincaré and Galilei groups are Lie groups, i.e., their elements form a continuous manifold. Associated with a Lie group is its Lie algebra, which describes the structure of the group in the vicinity of the identity element. When the elements of the group are interpreted as continuous finite transformations, the elements of the corresponding Lie algebra can be thought of as infinitesimal transformations near the identity, and are said to generate the former. The structure of a Lie group is thus encoded in the structure of its Lie algebra, and the latter is encoded in the commutation relations among its elements. Let $\{H, \mathbf{P}, \mathbf{J}, \mathbf{K}\}$ be the infinitesimal generators of time translations, space translations, spatial rotations, and inertial transformations (where $\mathbf{P}$, for instance, is short-hand for P_i, $i = 1, 2, 3$). For both the Poincaré and Galilei Lie algebras, these generators satisfy the following commutation relations:

$$\begin{array}{llll} [\mathbf{J},\mathbf{J}] = \mathbf{J} & [\mathbf{J},\mathbf{K}] = \mathbf{K} & [\mathbf{J},\mathbf{P}] = \mathbf{P} & [\mathbf{J},H] = 0 \\ [\mathbf{P},\mathbf{P}] = 0 & [\mathbf{P},H] = 0 & [\mathbf{K},H] = \mathbf{P} & \end{array} \tag{3.14}$$

(where, for instance, $[\mathbf{J},\mathbf{J}] = \mathbf{J}$ is short-hand for $[\mathcal{J}_i, \mathcal{J}_j] = i\varepsilon_{ijk}\mathcal{J}_k$). In addition to equation (3.14), the Poincaré Lie algebra is further characterized by

$$[\mathbf{K},\mathbf{K}] = -\mathbf{J} \qquad [\mathbf{K},\mathbf{P}] = H \tag{3.15}$$

while, in addition to equation (3.14), the Galilei Lie algebra is further characterized by

$$[\mathbf{K}, \mathbf{K}] = 0 \qquad [\mathbf{K}, \mathbf{P}] = 0 \tag{3.16}$$

One can obtain equation (3.16) from equation (3.15) by performing what Bacry and Lévy-Leblond (1969: 1609) call a "speed–space" contraction. The general idea is to construct a new Lie algebra by taking a suitable limit of the Poincaré Lie algebra. For a "speed–space" contraction, one defines a new set of generators $\{\mathbf{J}', H', \mathbf{P}', \mathbf{K}'\}$, where $\mathbf{J}' = \mathbf{J}$, $H' = H$, $\mathbf{P}' = \epsilon\mathbf{P}$, $\mathbf{K}' = \epsilon\mathbf{K}$, and $\epsilon \in \mathbb{R}$. The new generators are required to satisfy Lie bracket relations defined as the $\epsilon \to 0$ limit of the Poincaré Lie brackets. It's not hard to convince yourself that ϵ will drop out of all the expressions in equation (3.14); thus these will remain unchanged in the new Lie algebra. The only changes occur for the two expressions in equation (3.15). One has

$$\begin{aligned}[\mathbf{K}', \mathbf{K}'] &= [\epsilon\mathbf{K}, \epsilon\mathbf{K}] = \epsilon^2[\mathbf{K}, \mathbf{K}] = -\epsilon^2\mathbf{J} = -\epsilon^2\mathbf{J}'. \\ [\mathbf{K}', \mathbf{P}'] &= [\epsilon\mathbf{K}, \epsilon\mathbf{P}] = \epsilon^2[\mathbf{K}, \mathbf{P}] = \epsilon^2 H = \epsilon^2 H'.\end{aligned} \tag{3.17}$$

The $\epsilon \to 0$ limit of the right-hand-sides of both of these expressions then yields equation (3.16), and we identify the new contracted Lie algebra as belonging to the Galilei group. In general, the $\epsilon \to 0$ limit can be singular if one is not careful in correctly identifying the subset of generators with respect to which the contraction transformation is defined (i.e., the transformation that takes an initial generator $\mathbf{A}$ to a new generator $\mathbf{A}' = \epsilon\mathbf{A}$). It turns out that the limit is nonsingular if and only if the subset of generators that defines the contraction transformation forms a subgroup of the original group (Inonu and Wigner, 1953: 512–513). In the case of a speed–space contraction, the set $\{\mathbf{J}, H\}$ indeed forms a subgroup, $SO(3) \times \mathbb{R}$, of the Poincaré group.

Note that a speed–space contraction corresponds to taking two limits, namely, the limit of small velocities (the parameters associated with the boost generators $\mathbf{K}$) and the limit of small spatial translations (the parameters associated with the spatial translation generators $\mathbf{P}$). Another way to see how the Galilei group emerges from these limits is to consider their effects on Lorentz coordinate transformations. A Lorentz transformation is characterized by,

$$\begin{aligned} t \mapsto t' &= \gamma(t - (\mathbf{v} \cdot \mathbf{x})/c^2) \\ \mathbf{x} \mapsto \mathbf{x}' &= \gamma(\mathbf{x} - \mathbf{v}t) \end{aligned} \tag{3.18}$$

where $\gamma = (1 - |\mathbf{v}|^2/c^2)^{-1/2}$. In the low-velocity limit $|\mathbf{v}|/c \ll 1$, the factor γ goes to unity and equations (3.18) become,

$$\begin{aligned} t \mapsto t' &= t - (\mathbf{v} \cdot \mathbf{x})/c^2 \\ \mathbf{x} \mapsto \mathbf{x}' &= \mathbf{x} - \mathbf{v}t \end{aligned} \tag{3.19}$$

In the limit $|\mathbf{x}| \ll ct$ of small spatial translations, equations (3.19) become the (homogeneous) Galilean coordinate transformations:

$$\begin{aligned} t &\mapsto t' = t \\ \mathbf{x} &\mapsto \mathbf{x}' = \mathbf{x} - \mathbf{v}t \end{aligned} \tag{3.20}$$

Thus under a speed–space contraction, we assume that velocities are small compared with the speed of light, and in addition, we assume that timelike intervals are much greater than spacelike intervals. In the latter "ultra-timelike separation" limit, spacelike intervals become spatial intervals at equal times. Geometrically, taking both limits corresponds to flattening the light cones in Minkowski spacetime toward the spatial axes. This results in neo-Newtonian spacetime in which the Minkowskian distinction between timelike, lightlike, and spacelike worldlines has collapsed into just one type, namely, timelike worldlines.

The upshot of the discussion so far is that there is a well-defined relation between the Poincaré group and the Galilei group, based on a speed–space contraction. It is important to note that this relation induces a relation between irreducible representations of the Poincaré and Galilei groups.[18] On the other hand, it does not seem possible to extend this relation to one between the Poincaré group and symmetry groups of classical spacetimes other than the Galilei group. This is due to the facts that the contraction of a group must have the same dimension as the group, and the Galilei group is the only classical spacetime symmetry group with the same dimension (i.e., ten) as the Poincaré group.[19] Again, it might be worthwhile to point out that the NQFTs of physical interest are GQFTs, and that NQFTs based on classical spacetime symmetry groups other than the Galilei group may only be of interest in incorporating gravity consistently into the QFT framework (the idea being that the additional degrees of freedom associated with other classical spacetime symmetry groups, like the Maxwell group, encode the dynamical degrees of freedom of the gravitational field).

Let's consider the ramifications of the above discussion for explanations of why the Spin–Statistics and CPT theorems fail in NQFTs. There seem to be two relevant characteristics of the intertheoretic link between RQFTs and NQFTs represented by a speed–space contraction:

(a) A *speed–space* contraction transforms the Poincaré group into the Galilei group, and induces a transformation of irreducible representations of the former into irreducible representations of the latter.

(b) A *speed–space* contraction induces a transformation that takes spacelike intervals into spatial intervals at equal times. This, in turn, induces a transformation between relativistic locality constraints that are phrased in terms of spacelike intervals, and non-relativistic locality constraints that are phrased in terms of spatial intervals at equal times. In particular, under a speed–space contraction, the relativistic versions of LC, CD, *causality*, and *algebraic causality* transform into their non-relativistic versions.

Characteristics (a) and (b) can be thought of as underwriting a *kinematical* intertheoretic relation between RQFTs and GQFTs in each of the approaches discussed in Sections 1.2 and 3.2. This relation is kinematical in the sense that it maps the spatiotemporal structure associated with RQFTs into the spatiotemporal structure associated with GQFTs. Thus under a speed–space contraction,

(i) *Axiomatic approach*: In Table 3.1, axioms W2 (*Poincaré invariance*) and W3 (LC) for Wightman axiomatic RQFTs are transformed into axioms L2 (*Galilei invariance*) and L3 (*non-relativistic local commutativity*), respectively, for Lévy-Leblond axiomatic GQFTs.

(ii) *Weinberg's approach*: In Section 3.2.2, conditions (2) (relativistic invariance of the S-matrix) and (3) (*relativistic* CD) for Weinbergian RQFTs are transformed into conditions (2′) (non-relativistic invariance of the S-matrix) and (3′) (*non-relativistic* CD) for Weinbergian NQFTs.

(iii) *Lagrangian approach*: A restricted Lorentz invariant Lagrangian is transformed into a Lagrangian invariant under the Galilei group, and the (relativistic) *causality* constraint of Section 1.2.3 becomes the *non-relativistic causality* constraint of Section 3.2.3.

(iv) *Algebraic approach*: In Table 3.2., axioms HA2 (*Poincaré invariance*) and HA3 (*algebraic causality*) for Haag–Araki algebraic RQFTs are transformed into axioms 2 (*non-relativistic invariance*) and 3 (*non-relativistic algebraic causality*), respectively, for algebraic NQFTs.

3.3.3 A Kinematical Intertheoretic Relation

At this point, two observations should be made concerning the sense in which a speed–space contraction is an intertheoretic relation between *kinematical* features of RQFTs and GQFTs. The first involves the sense in which a speed–space contraction fails to encode all possible dynamical features of a given RQFT. The second observation points out that the restriction of the intertheoretic relation to purely kinematical features of RQFTs does not completely ignore some generic dynamical features of NQFTs.

With respect to the first observation, note that a speed–space contraction relation is intended to be an intertheoretic relation between what might be called theory frameworks, and not between individual theories. To the extent that a theory framework encodes the kinematics of a general class of theories without specifying particular dynamics, a speed–space contraction relation relates the purely kinematical framework of an RQFT to that of a GQFT. For a relation between specific theories, additional constraints may have to be imposed. Holland and Brown (2003), for instance, identify the following four constraints that must be imposed in order to describe the Galilean limit of Poincaré-invariant theories of the Maxwell and Dirac equations:

$$|\mathbf{v}|/c \ll 1,\ |\mathbf{x}| \ll ct,\ |\dot{\mathbf{x}}| \ll c,\ \nabla \gg (\mathbf{v}/c^2)(\partial/\partial t)$$

The first two constraints are the ingredients of a speed–space contraction, the third constraint refers to the velocity $\dot{\mathbf{x}}$ of an observed system (as opposed to the relative velocity $\mathbf{v}$ of an inertial frame), and the last constraint is needed to recover the correct Galilean transformation on the gradient operator. Holland and Brown point out that in general a speed–space contraction is insufficient to recover the non-relativistic limit of a given theory:

> It is an old result that, at the level of kinematics, the Galilean group is a limiting case of the Lorentz group (Inonu and Wigner, 1953). This in itself does not, however, constitute a general demonstration of the emergence of the expected Galilean relations in the non-relativistic limit of low velocities. To examine this, one must investigate the relation between the two symmetries in the context of specific dynamical theories. (Holland and Brown, 2003: 185)

This is consistent with the attitude adopted in this section. The concern of this section in articulating an intertheoretic relation between RQFTs and GQFTs is a purely kinematical concern that is independent of the dynamics of a particular theory, be it a particular RQFT or a particular GQFT. This is because the SSC and CPT invariance are purely kinematical properties in the sense that they can be derived in RQFTs at the level of kinematics alone: *regardless* of the dynamics that distinguishes individual RQFTs from one another, *all* RQFTs must possess these properties.

Given that a speed–space contraction underwrites an intertheoretic relation that is purely kinematical, there is still a sense in which it captures an important generic *dynamical* difference between RQFTs and GQFTs. This is essentially due to the fact that formulations of GQFTs, both pragmatist and purist, do not suffer the *existence problem*.

Recall from Section 1.4 that Haag's theorem is one way the *existence problem* for pragmatist approaches to RQFTs manifests itself. Haag's theorem indicates that, under reasonable assumptions, the Hilbert spaces for interacting and non-interacting states of an RQFT belong to unitarily inequivalent representations of the canonical (anti-) commutation relations, and this obstructs the definition of a (realistic) S-matrix as a unitary transformation between interacting particle states and non-interacting asymptotic particle states. Purist approaches to RQFTs take this to heart and attempt to construct interacting RQFT states directly from a model of a relevant set of axioms; but recall that the fact that there currently are no realistic models of such axioms is the purist's version of the *existence problem.* It turns out that the *existence problem,* in either pragmatist or purist forms, does not afflict GQFTs. On the one hand, pragmatist approaches to GQFTs do not suffer the consequences of Haag's theorem; on the other hand, there are interacting models of purist GQFT axioms.[20] The upshot of this is that the intertheoretic relation induced by a speed-space contraction can relate a *non-interacting* RQFT, on the one hand, to an *interacting* GQFT, on the other hand. In principle, this makes it possible to "push-down" essential properties of a non-interacting (or unrealistic interacting) RQFT, like CPT invariance and the SSC, to an interacting GQFT. Such a possibility will be considered further in Chapter 5 as a key characteristic of an adequate explanation of CPT invariance and the SSC. The remainder of this section provides a few more of the details of how GQFTs in particular, and NQFTs in general, avoid the consequences of Haag's theorem. This discussion follows that in Bain (2011: 104–105).

The version of Haag's theorem I would like to consider is what Earman and Fraser (2006: 313–314) term the Haag–Hall–Wightman (HHW) theorem. For two local fields ϕ_1, ϕ_2, the first part of the HHW theorem demonstrates that, under the assumptions,

(a) the fields belong to irreducible representations of the equal-time canonical commutation relations;

(b) there are unique Euclidean-invariant vacuum states $|0_1\rangle$, $|0_2\rangle$;

(c) there is a unitary transformation $V(t)$ that relates the fields at a given time;

then the vacuum states are constant multiples of each other: $c\,|0_1\rangle = V(t)\,|0_2\rangle$, $|c| = 1$. The second part of the HHW theorem demonstrates that, under the assumptions,

(d) (a)–(c);

(e) the fields and vacuum states are Poincaré-invariant;

then the first four vacuum expectation values of the two fields are equal. One can then demonstrate that if one of the fields is free, both are free (Earman and Fraser, 2006: 314). Thus if we insist that there are such things as *interacting relativistic* quantum fields, then in the first instance, we have to deny that the vacuum state of any such interacting field is a constant multiple of the corresponding free vacuum state. And, in the second instance, this entails that we must give up one or more of assumptions (a), (b), or (c). If one is wedded to the notion that a Fock space representation is an essential aspect of the formulation of a QFT, one may be loath to give up assumptions (a) and (b), in so far as both are necessary for a Fock space representation of a free field. One is thus compelled to reject assumption (c).[21]

The condition that the vacuum states of a free and an interacting field are multiples of each other is equivalent to the condition that the interaction does not polarize the vacuum. Vacuum polarization occurs when the Hamiltonian operator that describes the interacting field fails to annihilate the vacuum of the free field. Let $H = H_0 + H_I$ be an interacting Hamiltonian operator with free part H_0 and interaction part H_I. Suppose further that there is a free vacuum $|0_F\rangle$ such that $H_0\,|0_F\rangle = 0$. Then the interaction is said to polarize the vacuum just when $H\,|0_F\rangle \neq 0$. Vacuum polarization is conceptually problematic: the Hamiltonian operator encodes the energy of the system, so it should act on the state with zero energy (the vacuum state) and produce zero. In the first part of the HHW theorem, let H_1 and H_2 be Hamiltonian operators associated with the fields ϕ_1, ϕ_2, such that $H_1 |0_1\rangle = 0 = H_2 |0_2\rangle$. Then, since $c |0_1\rangle = V(t) |0_2\rangle$ implies that $H_2 |0_1\rangle = 0$, we can replace the condition that the vacuum states are constant multiples of each other with the condition that vacuum polarization does not occur. (And, obviously, if vacuum polarization does not occur, then the free and interacting vacuum states must be constant multiples of each other.)

This gives us two necessary conditions for the existence of interacting fields that are unitarily equivalent to free fields; namely,

(i) The interaction polarizes the vacuum; or

(ii) Poincaré invariance does not hold.

One can now argue that it is the absolute temporal structure of classical spacetimes that allows NQFTs in general to satisfy (ii) while denying the conceptually problematic (i).

This argument was suggested by Lévy-Leblond (1967: 160–161) to explain how interacting GQFTs avoid Haag's theorem. He points out that the generator of time-translations nowhere occurs on the right hand side of the Galilei Lie algebra, i.e., equations (3.14) and (3.16). Intuitively, time-translations are independent of the other generators, and this encodes the fact that the temporal metric is absolute in neo-Newtonian spacetime. One can now consider a representation of the generators on a state space that encodes the time-translation generator as the Hamiltonian operator H_0 of a free field. If we then construct a Hamiltonian operator $H = H_0 + H_I$ that consists of this free part and a part H_I that describes an interaction, then the "free" representation $(H_0, \mathbf{P}, \mathbf{K}, \mathbf{J}, M)$ will be unitarily equivalent to the "interacting" representation $(H, \mathbf{P}, \mathbf{K}, \mathbf{J}, M)$ in the sense of satisfying the same commutation relations. The only constraint is that the interaction term H_I be Galilei-invariant. And one can then show that if the free Hamiltonian annihilates a vacuum state, then so does the interacting Hamiltonian (Fraser, 2006: 40, footnote 23). Hence no vacuum polarization occurs.

Lévy-Leblond now asks us to contrast this with the situation arising in the case of the Poincaré group. The commutation relations that define its generators, equations (3.14) and (3.15), include one in which the generator of time-translations is "mixed up" with the generators of pure Lorentz boosts and space-translations: $[\mathbf{K}, \mathbf{P}] = H$. This encodes the fact that there is no independently occurring absolute temporal metric in Minkowski spacetime. Hence requiring that H_I be Poincaré-invariant will not guarantee that $H_0 + H_I$ will preserve the Lie bracket structure of the representation in which H_0 appears (Fraser, 2006: 41). In general, another structurally distinct representation of the Poincaré generators will have to be constructed for the interacting Hamiltonian $H_0 + H_I$. Thus if a free Hamiltonian annihilates a Poincaré-invariant vacuum state, this does not guarantee that an interacting Hamiltonian will do so, too.

The moral then is that the existence of interacting GQFTs that do not polarize the vacuum reflects the absolute temporal structure of neo-Newtonian spacetime.

One would expect that this way of avoiding Haag's theorem extends to all NQFTs, given that all have in common the classical spacetime structure associated with an absolute temporal metric. The key condition is that the generator of time translations be independent, as it were, of the other generators, and this will be the case for the symmetry group of any spacetime with an absolute temporal metric. This inference is given support by explicit examples of Fock space representations of interacting GQFTs (Lévy-Leblond, 1967), as well as Christian's (1997) Maxwell-invariant QFT of Newtonian gravity. The latter is an interacting NQFT that admits a Fock space representation and a total number operator (Christian, 1997: 4872). Thus one might infer that any NQFT will not run afoul of Haag's theorem.[22]

3.4 Summary

Why are CPT invariance and the SSC not derivable properties in NQFTs? The answer to this question depends on the specific approach to NQFTs one adopts. In the axiomatic approach (derived from the Wightman axiomatic approach in RQFTs), the answer involves noting two things: first, the complex version of the classical spacetime symmetry group associated with an axiomatic NQFT does not have a component connected to the identity that contains the PT transformation; and second, the vacuum state of an axiomatic NQFT is not separating. In Weinberg's approach, the answer involves noting that vacuum expectation values of time-ordered products of non-relativistic quantum fields are always invariant under the corresponding classical spacetime symmetry group (as long as the fields themselves are likewise invariant). In the Lagrangian approach, the answer involves noting the differences between hyperbolic and parabolic partial differential equations. Finally, in the algebraic approach, the answer can be phrased simply in terms of the fact that the vacuum state of an algebraic NQFT is not separating.

These explanations of the failure of CPT invariance and the SSC to be derivable properties in NQFTs have two things in common. First, they are restricted to the context of a particular approach to formulating NQFTs. This is important, in so far as one takes the conceptual and foundational differences between these approaches seriously. Second, they all involve more than a simple appeal to the failure of relativity. In the axiomatic, Weinberg, and Lagrangian approaches, an additional appeal to the failure of a relativistic locality constraint was made. In the algebraic approach, just the latter appeal was made.

RQFTs can be related to NQFTs by an intertheoretic relation induced by a speed–space contraction of the Poincaré group. This contraction involves both a low-velocity limit and an ultra-timelike separation limit in which spacelike intervals are small compared with timelike intervals. In recovering the Galilei group from the Poincaré group, these limits transform RLI into Galilei invariance, and the various relativistic locality constraints (LC, CD, *causality, algebraic causality*) into non-relativistic versions. An important aspect of this intertheoretic relation is that it transforms kinematical features of non-interacting (or at most unrealistic interacting) RQFTs into both kinematical and dynamical features of interacting GQFTs. This intertheoretic relation will be important in the discussion in Chapter 5 of what explains CPT invariance and the SSC.

NOTES

1. I follow Araki's (1999: 103–104) treatment of the Wightman axioms and adapt it, for the sake of comparison, to the treatment of the Lévy-Leblond axioms given in Lévy-Leblond (1967). In W1 and L1, the field operators should be defined as operator-valued distributions, and accommodations should be made for unbounded operators. These details will be glossed over in the following. Furthermore, in both cases, an additional axiom of asymptotic completeness, important for scattering theory, will not be needed in the subsequent discussion. Finally, I consider only the $m \neq 0$ case for simplicity. For other axiomatic approaches to GQFTs, see Dadashev (1985) and Puccini and Vucetich (2004a).
2. For the relativistic case, see Araki (1999: 103). For the Galilei case, see Lévy-Leblond (1967: 163).
3. *Cyclicity* of the vacuum, axiom (4ii), is the requirement that acting on the vacuum with operators defined on $\mathcal{H}$ (axiom (W4ii)), or within a mass sector of $\mathcal{H}$ (axiom (L4ii)) yields all states in $\mathcal{H}$ (*resp.* within the mass sector). In the axiomatic treatment, cyclicity guarantees that the fields form an irreducible representation of the equal time canonical commutation relations (Streater and Wightman, 1964: 101). This is a necessary condition for the construction of a Fock space representation.
4. There are two ways of obtaining expression (3.2). One can second-quantize the Schrödinger equation by first obtaining a solution expanded as a Fourier series (in the case of the Schrödinger equation, there is only one Fourier "positive energy" mode), and then promoting the expansion coefficient to an annihilation operator that acts on the Galilei-invariant multi-particle states of a Fock space. Alternatively (but equivalently), one can construct a Fock space of Galilei-invariant single-particle states as carriers of the scalar spin-0 irreducible representation of the Galilei group, define creation and annihilation operators on this Fock space, and then obtain equation (3.2) as the configuration space Fourier transformation of the annihilation operator.
5. The Pauli matrices are given by

$$\sigma_0 = \begin{pmatrix} 1 & 0 \\ 0 & 1 \end{pmatrix}, \quad \sigma_1 = \begin{pmatrix} 0 & 1 \\ 1 & 0 \end{pmatrix}, \quad \sigma_2 = \begin{pmatrix} 0 & -i \\ i & 0 \end{pmatrix}, \quad \sigma_3 = \begin{pmatrix} 1 & 0 \\ 0 & -1 \end{pmatrix}.$$

6. Under this correspondence, $\Lambda^{\mu}_{\nu} = (1/2)\mathrm{Tr}[\sigma^{\mu} A \sigma_{\nu} A^{\dagger}]$. From this one infers that only restricted Lorentz transformations enter into this correspondence

(i.e., Lorentz transformations for which $\Lambda_0^0 = (1/2)\mathrm{Tr}[AA^\dagger] > 0$, and that can be continuously connected to the identity). Moreover, to every such restricted Lorentz transformation, there correspond two $SL(2, \mathbb{C})$ matrices $A, -A$. Thus the map is 2:1, and in addition, it can be shown to be a homomorphism, i.e., it preserves the group product of $SL(2, \mathbb{C})$.

7. G can be decomposed as $G = SO(3, \mathbb{R}) \ltimes [(\mathbb{R}^3 \times \mathbb{R}^3) \ltimes \mathbb{R}^1]$, i.e., the semi-direct product of the group $[(\mathbb{R}^3 \times \mathbb{R}^3) \ltimes \mathbb{R}^1]$ (consisting of space translations, velocity boosts, and time translations) with $SO(3, \mathbb{R})$ (Lévy-Leblond, 1971: 229). This entails that $SO(3, \mathbb{R})$ is the quotient group $G/[(\mathbb{R}^3 \times \mathbb{R}^3) \ltimes \mathbb{R}^1]$, which means $SO(3, \mathbb{R})$ is homomorphic to G.
8. This leaves open the possibility that the T transformation is contained in the connected component of a *complex* classical symmetry group. But naively, simply complexifying a classical 4-vector will not mix up its spatial and temporal parts.
9. Technically, this assumes a particular way of identifying spatial regions with subalgebras of operators on $\mathcal{H}$, namely, what Halvorson (2001: 117) calls the *standard localization scheme*, which assigns to a spatial region $\mathcal{S}$, the relevant Cauchy data with support in $\mathcal{S}$ (intuitively, this Cauchy data is in 1–1 correspondence with those local field observables, viewed as solutions to the relevant field equations, with support in $\mathcal{S}$).
10. In general, cyclicity for *spatiotemporal* local algebras, relativistic or non-relativistic, is a property of *any* state that is analytic in the energy.
11. The partial differential equations (PDEs) of interest in non-relativistic and relativistic QFTs are of the parabolic form $u_t + Lu = 0$, and the hyperbolic form $u_{tt} + Lu = 0$, respectively, where L is a second order elliptic operator dependent on the spatial coordinates. These PDEs are obtained as the configuration space representation of the SC (defined explicitly on momentum space variables) and inherit the signature of the spacetime through the (inverse) Fourier transformation of the momentum space variables. The result is an elliptic PDE in Riemannian spacetimes, a hyperbolic PDE in Lorentzian spacetimes, and a parabolic PDE in classical spacetimes (see, e.g., McCabe, 2007: 41–43).
12. Normal-ordering is still necessary if we wish to identify the non-relativistic fields that appear in our local Lagrangian density as operator-valued distributions (since products of distributions at the same point are not necessarily well-defined).
13. In the sense that $\mathcal{J}\Omega = \Omega = \Delta\Omega$, $\Delta^{it}\mathfrak{R}\Delta^{-it} = \mathfrak{R}$, and $\mathcal{J}\mathfrak{R}\mathcal{J} = \mathfrak{R}'$, where $\mathfrak{R}'$ is the commutant of $\mathfrak{R}$.
14. So-called by Stachel (2003: 20). The origin of the cube and "$cG\hbar$-physics" are discussed in Gorelik and Frenkel (1994: 86–95). The version of the cube in Figure 3.1 follows the description found in Christian (1997).
15. This is the geometrized version of Newton's theory of gravity originally described by Elie Cartan (see, e.g., Bain, 2004).
16. This is a quantized version of Newton–Cartan gravity formulated by Christian (1997). Bain (2004) describes it as an NQFT that is invariant under an

extension of the Maxwell group (*Maxwell*) of Section 1.1. It identifies the degrees of freedom of a quantized Newtonian gravitational field with aspects of the curvature of classical Newton–Cartan spacetime.

17. An example of such a system is a charged particle moving in a region external to an operating solenoid. Quantization of this system produces the Aharonov–Bohm effect (see, e.g., Belot, 1998: 546).
18. See, e.g., Brennich (1975). There is a similar relation between a trivial extension of the Poincaré group and the extended Galilei group, and this likewise induces a relation between irreducible representations of the Poincaré group and irreducible representations of the extended Galilei group. The (trivially) extended Poincaré Lie algebra replaces equation (3.15) with $[\mathbf{K}, \mathbf{K}] = -\mathbf{J}$, $[\mathbf{K}, \mathbf{P}] = (M + H)$. The counterparts of these relations in the extended Galilei Lie algebra are $[\mathbf{K}, \mathbf{K}] = 0$, $[\mathbf{K}, \mathbf{P}] = M$.
19. The Poincaré group also admits a contraction with respect to a subgroup generated by $\{\mathbf{J}, \mathbf{P}\}$. Under this "speed–time" contraction, one makes the substitutions $H \to \epsilon H$, $\mathbf{K} \to \epsilon \mathbf{K}$ in equations (3.14) and (3.15), and then takes the $\epsilon \to 0$ limit (Bacry and Lévy-Leblond, 1968: 1609). The result is a ten-parameter group that Lévy-Leblond (1965) christened the Carroll group. A speed–time contraction combines a low-velocity limit $|\mathbf{v}|/c \ll 1$ with an "ultra-spacelike separation" limit $|\mathbf{x}| \gg ct$ (i.e., spacelike intervals are large compared with timelike intervals). Geometrically, a speed–time contraction flattens light cones in Minkowski spacetime toward the temporal axis (Lévy-Leblond, 1965: 3–4). This results in a spacetime in which the Minkowskian distinction between timelike, lightlike, and spacelike worldlines has collapsed into just one type, namely, spacelike worldlines. (The acausal nature of this spacetime reminded Lévy-Leblond (1965: 11) of *Alice in Wonderland*, hence the name "Carroll group.") From the point of view of Section 3.1, the Carroll group is not a classical spacetime symmetry group, since it fails to leave the absolute temporal metric of a classical spacetime invariant.
20. In particular, there are interacting GQFTs that describe important aspects of many condensed matter systems.
21. If one is not wedded to Fock space representations, one may give up assumption (b) and avoid Haag's theorem by inserting a cut-off into one's interacting theory and renormalize the fields. Alternatively, some authors have proposed giving up assumption (a) (Streater and Wightman, 1964: 101).
22. Note that Fraser (2006: 46) indicates that *non-relativistic local commutativity* (axiom L3 in Table 3.1) blocks the Streit–Emch version of Haag's theorem (which does not explicitly require Poincaré invariance). Lévy-Leblond (1967: 166) suggests the SC (axiom L4iii in Table 3.1) also plays a role. Since these two axioms are (arguably) shared by any NQFT, this further suggests that nothing unique to the Galilei group beyond what it shares in common with other non-relativistic spacetime symmetry groups does the work in avoiding the consequences of Haag's theorem.

4

Non-RQFT Derivations of CPT Invariance and the Spin–Statistics Connection

This chapter evaluates attempts to derive CPT invariance and the spin–statistics connection (SSC) outside the framework of relativistic quantum field theory. In particular, we will consider non-relativistic quantum mechanical attempts to derive the SSC, and relativistic non-quantum mechanical attempts to derive CPT invariance. The former are important to the extent that the SSC is essential in descriptions of many types of non-relativistic quantum mechanical systems, both field-theoretic and non-field-theoretic. The latter attempts are important in understanding the essential nature of CPT invariance; in particular whether it is essentially relativistic and/or essentially quantum mechanical.

Section 4.1 considers attempts at non-relativistic quantum mechanical derivations of the SSC. These attempts split into those that adopt the framework of finite-dimensional (i.e., non-field-theoretic) non-relativistic quantum mechanics (NQM), and fall under the general heading of "configuration space approaches," and an approach due to Sudarshan that adopts the framework of (field-theoretic) NQFT. Recent work in the philosophy of physics literature on the relation between permutation invariance and indistinguishability will shine light on the limitations of the former, while the discussion in Chapter 3 will indicate the limitations of the latter.

Section 4.2 considers attempts to derive CPT invariance in relativistic classical field theories (RCFTs) from both a pragmatist (Lagrangian) and a purist (algebraic) perspective. We shall see that the pragmatist attempt succeeds, whereas the purist attempt does not. The pragmatist derivation will indicate that the relation between C and PT in Lagrangian formulations of RCFTs is weaker than it is in Lagrangian formulations of RQFTs. In particular, in Lagrangian RCFTs, PT is the physically relevant symmetry, and CPT is derivative of it, whereas in Lagrangian RQFTs, CPT is the physically relevant symmetry, and PT is derivative of it.

4.1 Non-Relativistic Quantum Mechanical Derivations of the Spin–Statistics Connection

Attempts to derive the SSC in non-relativistic quantum mechanical systems can be divided into two camps: those for finite systems (i.e., systems described by NQM), and those for infinite systems (i.e., systems described by NQFTs). Section 4.1.1 reviews examples of the former while Section 4.1.2 reviews an example of the latter.

4.1.1 Configuration Space Approaches

The goal of what Papadopoulos et al. (2004: 206) refer to as "configuration space approaches" is to derive the SSC for a finite, many-particle non-relativistic quantum system by quantizing a topologically non-trivial classical configuration space of non-coinciding indistinguishable particles. These approaches originated in the work of Laidlaw and DeWitt (1974) and Leinaas and Myrheim (1977). One family of such approaches assumes the existence of antiparticles and CPT-type symmetries (Finkelstein and Rubinstein, 1968; Tscheuschner, 1989, 1990, 1991; Balachandran et al., 1993).[1] An alternative approach is Berry and Robbins (1997, 2000a, 2000b), who introduce position-dependent spin states. Attempts to extend their results include Harrison and Robbins (2000, 2004), Papadopoulos et al. (2004), Benavides and Reyes-Lega (2010), Papadopoulos and Reyes-Lega (2010), Reyes-Lega and Benavides (2010), and Reyes-Lega (2011). Similar approaches include Peshkin (2003a, 2003b, 2006) and Forte (2007).

To set the stage for a discussion of configuration space approaches, it will help to briefly review how statistics can be encoded in a quantum system. Recall from Section 1.1.2 that statistics can be encoded in a quantum field theory by imposing (anti-) commutation relations on either creation and annihilation operators that act on multi-particle states in a Fock space, or, more generally, on field operators (expressions (1.1) and (StLC), respectively). Both procedures entail that the corresponding multi-particle Fock space states (when they are available) are either symmetric or antisymmetric under a permutation of single-particle substates, and thus obey Bose–Einstein (BE) statistics or Fermi–Dirac (FD) statistics, respectively. The imposition of (anti-) commutation relations thus entails that quantum field theoretic states can only possess either BE statistics or FD statistics. This "Bose–Fermi alternative" condition is sometimes referred to as a "symmetrization postulate" that requires multi-particle Fock space states to be either symmetric or antisymmetric under permutation of their single-particle substates. Thus to impose (anti-) commutation relations on one's dynamical variables in a quantum field theory is to impose the symmetrization postulate by fiat. In principle, alternatives to BE and FD statistics are theoretically possible (in the sense of being consistent with a theory's equations of motion). Such alternatives are called parastatistics, and can be encoded in trilinear (as opposed to the standard bilinear) (anti-) commutation relations. More generally, the statistics possessed by a quantum state can be encoded in the irreducible representation of the group S_N of permutations of N objects under which it transforms. "Ordinary" statistics (BE or FD) correspond to the two one-dimensional (1-dim) representations of S_N, whereas parastatistics correspond to higher dimensional representations. On the other hand, all empirical evidence suggests that the quantum systems of interest possess either BE or FD statistics. Moreover, there is a sense in which parastatistics are conventional, insofar as a quantum system that possesses parastatistics can always be re-described in terms of a system that possesses either BE or FD statistics plus an internal gauge symmetry (Baker et al., 2014).

In non-relativistic (finite-dimensional) quantum mechanics, a similar story can be told, but one in which the roles of *particle identicality, particle indistinguishability*, and permutation invariance are typically stressed.[2] Particles are *identical* if they agree on all their state-independent (viz., "intrinsic") properties (i.e., mass, charge, spin, etc.). Typically, *particle indistinguishability* is cashed out in terms of *permutation invariance*: particles are said to be indistinguishable just when all observables commute with all particle permutations (Huggett and Imbo, 2009: 312). More precisely, let $H^N = H \otimes \ldots \otimes H$ be an N-fold tensor product of single-particle Hilbert spaces H, and let U_π be a unitary operator on H^N representing an element $\pi \in S_N$. U_π acts on an N-particle state $|\psi\rangle = |\psi\rangle_1 |\psi\rangle_2 \ldots |\psi\rangle_N \in H^N$ by permuting its single-particle Hilbert space indices:

$$U_\pi |\psi\rangle = |\psi\rangle_{\pi(1)} |\psi\rangle_{\pi(2)} \ldots |\psi\rangle_{\pi(N)}. \tag{4.1}$$

Permutation invariance (PI) requires that for any $|\psi\rangle \in H^N$ and any $\pi \in S_N$,

$$\langle \psi | A | \psi \rangle = \langle \psi | U_\pi^* A U_\pi | \psi \rangle \tag{PI}$$

for any observable represented by unitary operator A. In words: the expectation value of any observable A remains unchanged under single-particle permutations performed on a multi-particle state. PI entails that $[A, U_\pi] = 0$, for any observable A and any permutation operator U_π. As Baker et al. (2014: 7) point out, PI is a constraint on a theory's algebra of observables, but it induces the following constraint on elements $|\psi\rangle$, $|\psi'\rangle$ of H^N: for any observable A,

$$\langle \psi | A | \psi \rangle = \langle \psi' | A | \psi' \rangle \text{ *if and only if* } |\psi\rangle \text{ *and* } |\psi'\rangle \text{ *represent the same state.*} \tag{4.2}$$

The symmetrization postulate can now be stated as the requirement that each state of the system corresponds to a 1-dim subspace (i.e., ray) of H^N. Since the latter carry 1-dim irreducible representations of S_N, the symmetrization postulate, again, imposes the Bose–Fermi alternative by ruling out higher-dimensional representations of S_N (that encode parastatistics) by fiat.

I'd now like to turn to configuration space approaches to deriving the SSC in NQM. The initial aim of these approaches is to *derive* the symmetrization postulate, as opposed to imposing it by fiat. The strategy is to impose indistinguishability directly on the states of a classical configuration space, as opposed to imposing it on an algebra of quantum observables, as PI does above. The hope then is that quantizing a classical system of indistinguishable particles will induce the symmetrization postulate, in the first instance, and that this will then entail the SSC. The procedure thus involves the following steps:

1. One assumes a many-particle classical system in which the particles are indistinguishable and non-coinciding (i.e., impenetrable). The resulting many-particle configuration space turns out to be topologically non-trivial (in the sense that it is multiply-connected).
2. One then seeks to demonstrate that a quantum many-particle wavefunction defined on the classical configuration space of step 1 must satisfy the symmetrization postulate, i.e., it must either be symmetric or antisymmetric under permutation of single-particle substates.
3. Finally, one attempts to demonstrate that the many-particle wavefunction of step 2 possesses the SSC, i.e., it is either symmetric or antisymmetric, depending on its spin.

There are problems with each of these steps. Briefly, step 1 is motivated by associating classical particle indistinguishability with invariance under permutations; but most authors agree that the latter should be understood in terms of PI, namely, as a constraint imposed on the structure of the algebra of quantum observables, as opposed to a constraint on classical states.[3] Step 2 is flawed, as numerous commentators have pointed out. Simply put, PI by itself does not entail the symmetrization postulate, and an additional assumption in the form of a "single-valuedness" condition enforced on the wavefunction is suspect without further ado. Moreover, while step 2 can be accomplished for the special case of a *scalar* wavefunction for spinless particles, it has yet to be shown to hold in general for wave functions of arbitrary spin. More importantly, step 3 has only been accomplished under even more specialized conditions. These conditions will be reported at the end of this section. Before doing so, however, I would like to consider how the configuration space approach addresses steps 1 and 2 in a bit more detail.

Configuration space approaches begin by imposing indistinguishability on classical states in the following way (see, e.g., Morandi, 1992: 124–126; Landsman, 2013: 7). Let $Q = \mathbb{R}^d$ be a classical single-particle configuration space in d spatial dimensions (the points of Q represent the positions of the particle). The configuration space for a system of N identical but distinguishable particles, call it Q^N, consists of N copies of Q, i.e., $Q^N = \mathbb{R}^{Nd}$. For non-coinciding particles, this becomes $Q^N \backslash \Delta_N$, where we have excised from Q^N the set Δ_N of configurations in which two or more particles coincide; namely $\Delta_N = \{(x_1, \ldots, x_N) \in Q^N\colon x_i = x_j$ for at least one pair $(i,j), i \neq j\}$. Indistinguishability is now imposed by requiring that configurations that differ only by a permutation of particles represent the same state, i.e., one identifies the points $(x_1, \ldots, x_N)$ and $(x_{\pi(1)}, \ldots, x_{\pi(N)})$, where $\pi \in S_N$ is any permutation of the labels $\{1, \ldots, N\}$. The resulting collection of equivalence classes of so-identified points is the "reduced" configuration space Q_N of N indistinguishable (and non-coinciding) particles, and takes the form:

$$Q_N = \left(Q^N \backslash \Delta_N\right) / S_N. \tag{4.3}$$

One can now show that Q_N is multiply connected.[4] This suggests to some commentators that wavefunctions defined on Q_N are not "single-valued", in the sense that they are not well-defined globally.[5] To rectify this, typical accounts either implicitly or explicitly define wavefunctions on the universal covering space $\tilde{Q}_N$ of Q_N, defined by $Q_N = \tilde{Q}_N / \pi_1(Q_N)$, where $\pi_1(Q_N)$ is the first homotopy group of Q_N. Whereas Q_N is multiply-connected, $\tilde{Q}_N$ is simply-connected, thus wavefunctions defined on the latter are globally well-defined ("single-valued"). On the other hand, particle indistinguishability as encoded in the topological structure of Q_N has been "erased" on moving to $\tilde{Q}_N$. To (re)establish particle indistinguishability, one requires that, in the case of a *scalar* (i.e., spin-0) wavefunction $\tilde{\psi} : \tilde{Q}_N \to \mathbb{C}$,

$$\tilde{\psi}(g\tilde{q}) = a(g)\tilde{\psi}(\tilde{q}), \quad \tilde{q} \in \tilde{Q}_N, \quad g \in \pi_1(Q_N) \tag{4.4}$$

where the map $a : \pi_1(Q_N) \to U(1)$ is a 1-dim unitary representation of $\pi_1(Q_N)$ (i.e., a *character* of the group π_1) (Morandi, 1992: 119–120; Landsman, 2013: 8). One can then show that $\pi_1(Q_N) = S_N$, for $d \geq 3$.[6] Expression (4.4) then entails that for $d \geq 3$, the wavefunction transforms under a 1-dim representation of the permutation group, of which there are just two: the "symmetric" $a = 1$ representation, and the "antisymmetric" $a = \pm 1$ representation. For $d \geq 3$, the wavefunction must thus satisfy,

$$\tilde{\psi} \to a(g)\tilde{\psi} = \begin{cases} \tilde{\psi} \text{ if } a = 1 \text{ (Bose statistics).} \\ \pm\tilde{\psi} \text{ if } a = \pm 1 \text{ (Fermi statistics).} \end{cases} \tag{4.5}$$

Hence we have derived the symmetrization postulate for *scalar* (i.e., spin-0) quantum mechanical wavefunctions in $d \geq 3$.

As Landsman (2013: 8) points out, this derivation is misleading in at least two ways. First, the "single-valuedness" condition on the wavefunction that motivates the move to the universal covering space $\tilde{Q}_N$ is vague without further ado. Second, and perhaps more importantly, the restriction to *scalar* wavefunctions obscures the understanding of PI as a constraint on the algebra of quantum observables. In particular, it suggests, via expression (4.4), that PI requires that two unit vectors represent the same state if and only if they differ by a phase, whereas PI only entails the constraint on states given by expression (4.2). More generally, according to Landsman (2013: 8), "since all separable Hilbert spaces are isomorphic, particular realizations of states as wavefunctions are only meaningful in connection with some action of observables," and, without further ado, configuration space approaches fail to take this advice in establishing expression (4.5).

One way to address these concerns is to resist the move to the universal covering space $\tilde{Q}_N$, and rather adopt the multiply-connected Q_N as one's physical configuration space. To address concerns over "single-valuedness" (i.e., global well-definedness), this "intrinsic" approach defines wavefunctions as cross-sections of an appropriate vector bundle over Q_N.[7] This allows wavefunctions to be defined in a constraint-free manner. Intuitively, typical approaches that adopt the simply-connected $\tilde{Q}_N$ sacrifice particle indistinguishability at the level of the configuration space, and then reassert it as an additional constraint on the wavefunction. Call such approaches "extrinsic." The intrinsic approach, on the other hand, which adopts the multiply-connected Q_N, asserts particle indistinguishability at the level of the configuration space, and thus has no need to assert it as an additional suspect constraint on the wavefunction, provided the latter is defined in a global, intrinsic way. Moreover, a focus on the multiply-connected Q_N allows the intrinsic approach to adopt standard techniques associated with the quantization of multiply-connected phase spaces, and this allows one to take the structure of the algebra of observables into account.[8] These quantization techniques also make clear that the intrinsic approach does not employ a non-standard formulation of quantum mechanics, which is a common complaint of critics of extrinsic approaches (e.g., Allen and Mondragon, 2003). Finally, as Bourdeau and Sorkin 1992, pg. 687) suggest, "... the intrinsic approach is more natural in that it avoids the introduction of 'gauge variables' (the [particle] labels) which only have to be eliminated at a later stage by imposing symmetry conditions on [the wavefunction]." In the intrinsic approach, the derivation of the symmetrization postulate for scalar wavefunctions amounts to the observation that there are only two complex line bundles over Q_N, the cross-sections of which correspond to symmetric and antisymmetric wavefunctions.

Of course, from the point of view of this chapter, one would like to go beyond a derivation of the symmetrization postulate and recover the SSC. There are just a handful of results to report in this context. Peshkin (2003a, 2003b, 2006) has shown that, within the extrinsic configuration space approach, the assumption of continuity of the wavefunction, in addition to the single-valuedness constraint, suffices to exclude Fermi–Dirac statistics for spin-0 bosons. Kuckert (2004) has shown that the SSC can be characterized in terms of the unitary equivalence between angular momentum operators for single-particle and two-particle systems when restricted to appropriate domains. In particular, Kuckert (2004: 50) shows that a 2-dim non-relativistic two-particle system possesses the SSC if and only if there is a unitary operator U such that

$$j = 2U\mathcal{J}U^* \qquad (4.6)$$

where $\mathcal{J}$ is the total angular momentum operator (spin plus orbital angular momentum) of a single-particle state, and j is the total angular momentum operator of the corresponding two-particle state in its center of mass frame.[9] In three spatial dimensions, the SSC holds for a non-relativistic two-particle system if and only if

$$j_z = 2U\mathcal{J}_z U^* \tag{4.7}$$

holds (where j_z and $\mathcal{J}_z$ are the z-components of j and $\mathcal{J}$) when restricted to the subspace of states that are even under the z-parity operator (i.e., all two-particle states ψ such that $P_z\psi = \psi$, where $P_z : (z, y, z) \mapsto (z, y, -z)$) (Kuckert, 2004: 51). These results are suggestive of a general link between the intertwining relations among the angular momentum operator algebras of finite particle systems and the SSC. Reyes-Lega (2011: 13) suggests that a reformulation of Kuckert's results in the intrinsic configuration space approach would be a next step toward the formulation of a general spin–statistics theorem for NQM, but currently more work remains to be done.

The upshot of the discussion so far is that a proof of the spin–statistics theorem in NQM with finite degrees of freedom has yet to be constructed. This is not because the standard configuration space approach to such a proof is inherently flawed, as some critics have suggested, for one can adopt a rigorous intrinsic version of the approach that addresses critical concerns. The problem is just that within this more rigorous approach, a definitive general result has not yet been established.

4.1.2 Sudarshan's Approach

In contrast to configuration space approaches, which attempt to derive the SSC in the context of a non-relativistic quantum mechanical system with a finite number of degrees of freedom (i.e., NQM), an approach due to Sudarshan (1968) attempts to derive the SSC in the context of the Lagrangian approach to non-relativistic quantum field theories (NQFTs). This approach is based on a version of the spin–statistics theorem for RQFTs due to Schwinger (1951, 1958), and is reviewed in Duck and Sudarshan (1997, 1998), Shaji and Sudarshan (2003), Sudarshan and Duck (2003), Sudarshan and Shaji (2004), and Shaji (2009).[10] It demonstrates that imposing $SU(2)$ invariance on a sufficiently generic Lagrangian functional of Hermitian fields is sufficient to derive the SSC.

In a bit more detail, Sudarshan's approach starts with the kinematic (i.e., interaction-free) part $\mathcal{L}_{kin}$ of a Lagrangian density, the initial assumption being that the SSC is a purely kinematic property.[11] One then requires the general form of $\mathcal{L}_{kin}$ to satisfy the following conditions (Duck and Sudarshan, 1998: 295; Shaji, 2009: 3):

(1) *Rotational invariance.* $\mathcal{L}_{kin}$ is an $SU(2)$ scalar and is a functional of fields ξ which are finite-dimensional representations of $SU(2)$.

(2) *Hermiticity.* The fields are expressed in the "Hermitian field basis:" $\chi = \chi^\dagger$.

(3) $\mathcal{L}_{kin}$ is at most linear in first derivatives of the fields.
(4) $\mathcal{L}_{kin}$ is bilinear in the fields.

In condition (1), $SU(2)$ is the covering group of $SO(3)$, the group of real rotations. Condition (2) can initially be understood as requiring the fields to be expressed in the $SU(2)$ 2-spinor formalism; however, as will be discussed below, there is a bit more to this condition than just this. The 2-spinor formalism is based on the fact that the carrying space of representations of $SU(2)$ is a 2-dim complex vector space S endowed with an antisymmetric metric. Elements of S are spin-1/2 two-component spinors (these belong to the fundamental representation of $SU(2)$; they are the $SU(2)$ equivalents of the Weyl spinor representations of $SL(2, \mathbb{C})$). A 2-spinor algebra can be constructed in which a spin-n representation of $SU(2)$ (i.e., a spin-n $SU(2)$-invariant field) can be expressed as a tensor product of an even or odd number of $SU(2)$ 2-spinors, depending on whether n is an integer or a half-integer. Condition (2), understood initially in this way, in addition requires that the $SU(2)$ 2-spinor basis be Hermitian.

A Lagrangian density that possesses properties (1)–(4) takes the following general form (Duck and Sudarshan, 1998: 295):

$$\mathcal{L}_{kin} = \sum_{rs} \left\{ \frac{i}{2}(\chi_r \dot{\chi}_s - \dot{\chi}_r \chi_s) K^0_{rs} - \frac{i}{2} \sum_{j=1}^{3} (\chi_r \nabla_j \chi_s - \nabla_j \chi_r \chi_s) K^j_{rs} - \chi_r \chi_s M_{rs} \right\} \quad (4.8a)$$

or

$$\mathcal{L}_{kin} = \sum_{rs} \chi_r \Lambda_{rs} \chi_s \quad (4.8b)$$

where $\Lambda_{rs} = \frac{i}{2}(K^0_{rs} \overleftrightarrow{\partial}_t - K^j_{rs} \overleftrightarrow{\partial}_j - M_{rs})$, with $\overleftrightarrow{\partial} = \overrightarrow{\partial} - \overleftarrow{\partial}$, and K and M, and thus Λ, are numerical matrices that depend on spin indices r, s.[12] One now claims that the Lagrangian densities for both relativistic and non-relativistic fields can be written in the form of equation (4.8). In the case of relativistic fields, the (interaction-free) spin-1/2 Dirac equation is a first-order differential equation, and the corresponding Lagrangian density takes the explicit form of (4.8) for spin indices $r, s = 1 \ldots 4$ and the K matrix corresponding to the Dirac matrices. The Euler–Lagrange equations for relativistic spin-0 and spin-1 fields are second-order differential equations, but these can be rewritten as systems of first-order equations with corresponding Lagrangian densities of the form of equation (4.8), at the expense of additional components for the fields (see, e.g., Sudarshan and Duck, 2003: 647).

In the case of non-relativistic fields that satisfy the Schrödinger equation, Sudarshan (1968: 291) and Sudarshan and Duck (2003: 648) demonstrate that the corresponding Lagrangian density takes the form of equation (4.8) under condition (2). This demonstration indicates that the "Hermitian field basis" of condition (2) is not quite equivalent to expressibility in the 2-spinor formalism, as I initially assumed above. In particular, Sudarshan (1968) originally formulated condition (2) in terms of what he referred to as the "S-principle," which requires fields $\phi(x) = \chi(x) + \chi'(x)$ to be decomposable into negative $\chi(x)$ and positive $\chi'(x)$ frequency parts, and the action to be invariant under exchange of $\chi(x)$ and $\chi'(-x)$. This is equivalent to what is referred to as "crossing symmetry" (see, e.g., Weinberg, 1995: 269); informally, this requires that fields be expressible in terms of linear combinations of creation and annihilation operators associated with equal numbers of particles and antiparticles. Thus to demonstrate that the Lagrangian density for a non-relativistic Schrödinger field takes the form of equation (4.8), Sudarshan and Duck (2003: 648) require that the field ψ and its Hermitian conjugate be expressible as sums $\psi = \phi + i\chi$, $\psi^\dagger = \phi^T - i\chi^T$, involving Hermitian "basis" fields ϕ, χ and their transposes. In the following, then, I will take condition (2) to require that the fields be expressed as $SU(2)$ 2-spinors *and* satisfy crossing symmetry.

Given the general form (4.8), the derivation of the SSC involves two steps:

(I) One first demonstrates that the matrix Λ_{rs} is symmetric if and only if the fields that appear in it possess half-integer spin, and antisymmetric if and only if the corresponding fields possess integer spin.[13]

(II) One then demonstrates that the matrix Λ_{rs} is symmetric if and only if the fields that appear in it anticommute, and antisymmetric if and only if the corresponding fields commute.[14]

Steps (I) and (II) then entail that the fields that occur in $\mathcal{L}_{kin}$ are integer-spin if and only if they commute, and are half-integer spin if and only if they anticommute, which establishes that they possess the SSC.

This approach to the SSC is expressed by Sudarshan's (1968: 291) "fundamental theorem:" "In all rotationally invariant quantum field theories which maintain the symmetry between emission and absorption processes [i.e., crossing symmetry], Bose fields must have integral spin and Fermi fields must have half-integral spin." Note that relativistic fields are typically taken to satisfy crossing symmetry to the extent that they are solutions to hyperbolic partial differential equations that have both positive and negative energy solutions. The imposition of crossing symmetry is then associated with the introduction of equal numbers of antiparticles to particles, in order to be able to interpret negative energy solutions as positive energy antiparticle solutions, as discussed in Section 3.2.1. Non-relativistic fields, on the other hand, are solutions to parabolic partial differential equations, which only possess positive energy solutions. Hence crossing symmetry is not necessary to maintain a positive energy interpretation of these solutions. In fact, one can show explicitly that imposing crossing symmetry on a massive Galilei-invariant field is consistent with it possessing either BE or FD statistics.

To see this, recall from Section 3.2.1 that one may express a massive free field that satisfies the Schrödinger equation as a linear combination of creation and annihilation operators for particles of mass m and $-m$, respectively, and we may interpret the latter as antiparticles of mass m. Taking the spin-0 case for simplicity, this is expression (3.3) in Chapter 3:

$$\Phi(\mathbf{x}, t) = (2\pi)^{-3/2} \int d^3p \left[\xi e^{(-iEt+i\mathbf{p}\cdot\mathbf{x})} A(\mathbf{p}) + \eta e^{(iEt-i\mathbf{p}\cdot\mathbf{x})} B^\dagger(\mathbf{p})\right]$$

One can then show that this field satisfies expression (3.4) of Chapter 3:

$$\left[\Phi(\mathbf{x}, t), \Phi^\dagger(\mathbf{y}, t)\right]_\pm = \left(|\xi|^2 \pm |\eta|^2\right) \delta^{(3)}(\mathbf{x} - \mathbf{y})$$

Again, recall that this expression vanishes everywhere except for $\mathbf{x} = \mathbf{y}$, regardless of the values of ξ and η. In particular, it vanishes even in the absence of antiparticles ($\eta = 0$), and even if we require equal numbers $|\xi| = |\eta|$ of particles and antiparticles (i.e., crossing symmetry). Thus a spin-0 non-relativistic field can either commute (i.e., satisfy BE statistics) or anticommute (i.e., satisfy FD statistics) at equal times, even if it possesses crossing symmetry, as Puccini and Vucetich (2005: 1) point out.

Puccini and Vucetich (2004b) furthermore point out that condition (2) in Sudarshan's approach entails that the fields that appear in expression (4.8) must be Hermitian, and massive Galilei-invariant quantum fields cannot be Hermitian. Recall from Section 3.1 that projective irreducible representations of the Galilei group G correspond to non-projective irreducible representations of $\tilde{G}^*$, the central extension of the universal covering of G. This central extension involves a mass operator that induces a mass-dependent phase in the transformation rule for Galilei-invariant fields. In particular, as noted in Section 3.2.1, a massive Galilei-invariant quantum field Φ_σ with spin index σ transforms under a unitary representation $U(g)$ of $\tilde{G}^*$ via

$$U(g)\Phi_\sigma(\mathbf{x}, t)U(g)^{-1} = \exp\left[im\gamma(g; \mathbf{x}, t)\right] \sum D^{(s)}_{\sigma\sigma'}\left(R^{-1}\right) \Phi_{\sigma'}\left(\mathbf{x}', t'\right) \tag{4.9}$$

where $g \in \tilde{G}^*$, $\gamma(g; \mathbf{x}, t) = \frac{1}{2}\mathbf{v}^2 t + \mathbf{v} \cdot R\mathbf{x}$, $\mathbf{x}' = R\mathbf{x} + \mathbf{v}t + \mathbf{a}$, $t' = t + b$, and $D^{(s)}_{\sigma\sigma'}(R)$, $\sigma = -s, \ldots, s$, is the $(2s + 1)$-dim unitary matrix representation of $R \in SU(2)$. By taking the Hermitian conjugate of both sides of equation (4.9), one obtains the transformation rule for the Hermitian conjugated field $\Phi^\dagger_\sigma(\mathbf{x}, t)$ (see, e.g., Puccini and Vucetich, 2004b: 5):

$$U(g)\Phi^\dagger_\sigma(\mathbf{x}, t)U(g)^{-1} = \exp\left[-im\gamma(g; \mathbf{x}, t)\right] \sum D^{(s)}_{\sigma'\sigma}(R)\Phi^\dagger_{\sigma'}(\mathbf{x}', t') \tag{4.10}$$

Thus Φ_σ transforms as a field of mass m, whereas its Hermitian conjugate transforms as a field of mass $-m$, which implies that massive Galilei-invariant fields cannot be Hermitian.[15]

The upshot of this discussion is that, while Sudarshan's approach to the spin–statistics theorem is applicable to relativistic quantum fields that possess crossing symmetry, at best it applies to massless Galilei-invariant quantum fields.

4.2 Relativistic Classical Mechanical Derivations of CPT Invariance

This section reviews the extent to which CPT invariance can be derived in relativistic classical field theories (RCFTs). Section 4.2.1 considers classical derivations in the Lagrangian approach, while Section 4.2.2 considers the prospects for such derivations in alternative approaches.

4.2.1 The Lagrangian Approach

Bell (1955) indicates how PT invariance can be derived in the context of classical relativistic fields. His result has been extended by Greaves and Thomas (2014) to a general theorem that encompasses both classical and quantum field theories and PT and CPT invariance. The goal of this section is to unpack these proofs, considered as examples within the Lagrangian approach to RCFTs and RQFTs.

Bell's (1955) classical PT theorem states the following:[16]

> In a classical field theory, suppose (a) the fields transform as restricted Lorentz invariant tensors (i.e., integer-spin representations of $L_+^\uparrow$); and (b) the dynamical equations are partial differential equations that express the vanishing of a local polynomial combination of the fields and their derivatives. Then the fields are invariant under the proper Lorentz group L_+ (that includes PT transformations).

In the proof, Bell (1955: 494) encodes assumption (b) in the form of a tensor $G_{\alpha\beta\ldots}$ that is a polynomial $\mathcal{F}\left[T_{\mu\nu\ldots}(x)\right]$ of a finite number of tensor fields $T_{\mu\nu\ldots}(x)$; schematically, $\mathcal{F}\left[T_{\mu\nu\ldots}(x)\right] = G_{\alpha\beta\ldots}(x)$. Under a restricted Lorentz transformation $\Lambda_{\mu\nu} \in L_+^\uparrow$,

$$\mathcal{F}\left[\Lambda_{\mu\mu'}\Lambda_{\nu\nu'}\ldots T_{\mu'\nu'\ldots}(x)\right] = \Lambda_{\alpha\alpha'}\Lambda_{\beta\beta'}\ldots G_{\alpha'\beta'\ldots}(x) \tag{4.11}$$

This holds for the particular choice

$$\Lambda_{11} = \Lambda_{22} = 1,\ \Lambda_{33} = \Lambda_{44} = \cos i\theta,\ \Lambda_{34} = -\Lambda_{43} = \sin i\theta,\ \Lambda_{\mu\nu} = 0 \text{ otherwise,} \tag{4.12}$$

where θ is real (one can check that $\Lambda_{44} = \cosh\theta > 0$ for real θ, and $\det\Lambda = 1$). Upon substituting equation (4.12) into equation (4.11) and expanding both sides as power series in θ, one can equate the coefficients on both sides. This depends explicitly on the fact that $\mathcal{F}$ is a polynomial in the fields. A theorem in complex analysis then entails that equation (4.11) holds not only for real θ but also for complex θ.[17] For the choice $\theta = i\pi$, we have $\Lambda_{\mu\nu} = -\delta_{\mu\nu}$, which is a PT transformation. Bell then easily establishes that $\mathcal{F}$ is PT invariant.[18]

Bell's classical PT theorem adopts the same strategy as the Wightman axiomatic CPT theorem for RQFTs (Section 1.2.1); both demonstrate that invariance under $L_+^\uparrow$, combined with additional assumptions, entails invariance under $L_+(\mathbb{C})$ (the complex proper Lorentz group), and hence PT invariance (of course the Wightman axiomatic proof then goes on to establish CPT invariance). In the Wightman axiomatic proof, the additional assumptions were restricted Lorentz invariance (RLI) and the *spectrum condition* (SC). In Bell's proof, the additional assumptions are RLI and the restriction to tensor fields in assumption (a), and "polynomiality," as expressed in assumption (b). Arguably, it is the latter that plays the crucial role of allowing the extension from invariance under $L_+^\uparrow$ to invariance under $L_+(\mathbb{C})$. Thus polynomiality plays the role in Bell's PT theorem that the SC plays in the Wightman axiomatic CPT theorem.

I'd now like to review how Greaves and Thomas (2014) have extended Bell's result to a more general theorem, one instance of which is what they refer to as a classical CPT theorem. A bit of preliminary exposition will first be needed; in particular, we will need to understand the sense in which a classical field theory can be said to possess a charge conjugation symmetry, since the latter is supposed to map particle states to antiparticle states in RQFTs, and this distinction does not explicitly occur in classical theories. Note first that, in the Lagrangian approach, a classical field theory is a theory encoded in a Lagrangian density $\mathcal{L}[\phi]$ that is a functional of fields $\phi : M \to V$ that take the form of maps from a spacetime manifold M to a vector space V of field values (the latter can be scalar, vector, spinor, tensor, etc.). Second, recall that a conserved charge is associated with an internal symmetry group $\mathcal{G}$, and $\mathcal{G}$ is required to act (irreducibly) on the space of field values V. After Wallace (2009: 218), we can say that a classical Lagrangian field theory has a *conjugation symmetry* associated with an internal symmetry group $\mathcal{G}$, if V admits a complex structure $\mathcal{J}$, and there is a symmetry C of the theory such that $C^2 = 1$ and $C\mathcal{J} = -\mathcal{J}C$.[19]

In addition to the "classical" conjugation symmetry C, Wallace (2009: 218) also identifies a second type of charge conjugation, what he refers to as "*Q*-conjugation." This type of symmetry arises from field quantization, which involves the construction of a Fock space of multi-particle states from the properties of the solution space $\mathcal{S}$ of a relativistic classical field equation. In particular, one decomposes the *complexified* solution space $\mathcal{S}^C$ into positive and negative frequency subspaces, $\mathcal{S}^C = \mathcal{S}_+^C \oplus \mathcal{S}_-^C$, based on a foliation of spacetime into instantaneous 3-spaces. One then throws out the negative frequency subspace and norm-completes the positive frequency subspace to form a single-particle Hilbert space $\mathcal{H}_{1\mathrm{P}}$, which then serves as the basis for the construction of a Fock space of multi-particle states. Furthermore, as Wallace (2009: 216–217) explains, if the complexification of the action of $\mathcal{G}$ on V is reducible, then $\mathcal{H}_{1\mathrm{P}}$ decomposes into particle and antiparticle subspaces.[20] Wallace's *Q*-conjugation is defined as conjugation with respect to the complex structure of $\mathcal{S}^C$.[21] It is distinguished from classical charge conjugation in terms of the following: classical charge conjugation C is a symmetry of $\mathcal{H}_{1\mathrm{P}}$ insofar as it leaves the positive frequency subspace $\mathcal{S}_+^C$ invariant, and maps the particle subspace of $\mathcal{H}_{1\mathrm{P}}$ into the antiparticle subspace, and vice versa (when such a decomposition of $\mathcal{H}_{1P}$ is available).[22] *Q*-conjugation, on the other hand, is not a symmetry of $\mathcal{H}_{1P}$, insofar as it maps the positive frequency subspace to the negative frequency subspace, and vice versa.

We now have accumulated enough baggage to be able to interpret Greaves and Thomas's (2014: 58) general PT/CPT theorem:

> If a field theory is (a) *formal,* (b) possesses the SSC, and (c) is invariant under (the universal covering of) the restricted Lorentz group, then it is invariant under the composition $C_\$ \circ$ PT of a generalized charge conjugation transformation $C_\$$ with a PT transformation if and only if it is \$-Hermitian.

Condition (a) is spelled out in terms of Greaves and Thomas's (2014: 48) particular formulation of an abstract field theory; in the context of the Lagrangian approach, it requires the Lagrangian density to take the form of a polynomial in the fields and their spacetime derivatives. The generalized charge conjugation $C_\$$ takes either the form of Wallace's classical charge conjugation, which Greaves and Thomas label as $C_\#$, or Wallace's *Q*-conjugation, which Greaves and Thomas label as C_*. Finally, \$-Hermiticity is invariance under the composition $S \circ C_\$$, where S implements a reversal of order in a product of fields.

Let's first see how the General PT/CPT theorem reproduces the Lagrangian CPT theorem for RQFTs outlined in Section 1.2.3. Recall that the latter requires a local, Hermitian, normal-ordered Lagrangian density that is a functional of fields that satisfy RLI and possess the SSC. Locality of the Lagrangian density requires that it be constructed from polynomials in the fields and their derivatives evaluated at the same spacetime point, and normal-ordering requires positive frequency field modes (corresponding to Fock space creation operators) be ordered to the right of negative frequency field modes (corresponding to Fock space annihilation operators). Normal-ordering is sufficient to ensure that the products of fields evaluated at the same point are well-defined,[23] and this ensures that the Bose–Fermi alternative is well-defined (i.e., that the commutator/anticommutator of fields is well-defined). In the General PT/CPT theorem, condition (a) ensures locality of the Lagrangian density, and Greaves and Thomas (2014: 57) allow for normal-ordering to ensure condition (b) is well-defined. The "quantum" Lagrangian CPT theorem is then extracted from the General theorem under the condition that $C_\$$ be C_*. \$-Hermiticity then becomes $*$-Hermiticity, i.e., the composition $S \circ C_*$. This amounts to exchanging the positive and negative frequency field modes while reordering products of fields, which "... corresponds exactly to Hermitian conjugation of operators in QFT" (Greaves and Thomas, 2014: 54).

Bell's classical PT theorem can be extracted from the general theorem under the condition that $C_\$$ be the identity.[24] \$-Hermiticty then reduces to S, i.e., a reordering of products of fields. Greaves and Thomas (2014: 58) also extract a classical CPT theorem from the general theorem under the condition that $C_\$$ be $C_\#$. \$-Hermiticity then becomes #-Hermiticity, i.e., the composition $S \circ C_\#$.

This amounts to exchanging the particle and antiparticle subspaces of the theory ($C_{\#}$) while reordering products of fields (S). While the classical PT theorem seems physically reasonable, the classical CPT theorem does not. Intuitively, while C can be defined for a classical theory, its physical significance is questionable since classical theories have no need for a matter/antimatter distinction. This places doubt on the physical significance of #-Hermiticity; indeed, according to Greaves and Thomas (2014: pg. 55) "... there is no general reason it should be met."

In addition to a classical PT theorem, and quantum and classical CPT theorems, Greaves and Thomas (2014: 58) extract a quantum PT theorem from the general theorem under the condition that $C_{\$}$ be the composition $C_{*} \circ C_{\#}$. However, for $*$-Hermitian RQFTs, this simply states the corollary to the quantum CPT theorem that "... given the background assumptions required for the full CPT theorem, C-invariance is equivalent to PT-invariance." As Swanson (2014: 98) observes, the relations among these four versions of the General theorem suggest a structural difference between the classical and quantum theorems. The classical CPT theorem is derivative of the classical PT theorem, which of the two seems to be the more fundamental result: the classical CPT theorem is obtained from the classical PT theorem by requiring the theory to be $C_{\#}$-invariant. On the other hand, the quantum PT theorem is derivative of the quantum CPT theorem, in the sense that the former is a corollary of the latter. As Swanson notes, the classical theorems together entail that a classical field theory is CPT invariant just when it is C-invariant, whereas the quantum theorems allow a CPT-invariant quantum field theory to violate C-invariance. According to Swanson (2014: 98), "... there is interesting interaction between spatiotemporal and charge structure in the quantum theory which is entirely absent in the classical version." Thus in the CPT theorem for Lagrangian RCFTs, PT is the physically relevant symmetry, and C comes along for the ride as a physically questionable symmetry. In the CPT theorem for Lagrangian RQFTs, CPT is the physically relevant symmetry, and PT comes along for the ride.

4.2.2 Prospects for Alternative Approaches

The Lagrangian approach to the CPT theorem for RCFTs shares aspects of both the Lagrangian and Wightman axiomatic approaches to the CPT theorem for RQFTs. On the one hand, in both the classical and quantum Lagrangian approaches, fields are taken to be the basic objects, and these are required to appear in a suitably constrained Lagrangian density. On the other hand, in both the classical Lagrangian approach and the Wightman axiomatic approach, the basic strategy is to extend RLI of the fields to invariance under the complex proper Lorentz group. In the classical Lagrangian approach, this is done via the assumption of polynomiality, whereas in the Wightman axiomatic approach, it is done via the SC. This suggests that the classical analogue of the Wightman axiomatic approach to the CPT theorem can be found in the classical Lagrangian approach. Indeed, it is difficult to imagine an approach to RCFTs that more closely mimics the Wightman approach to RQFTs. The basic objects of the latter are vacuum expectation values of products of fields, for which there is no classical analogue. Similar remarks hold for a classical analogue for RCFTs of Weinberg's approach to RQFTs. While it is possible to construct an S-matrix for classical scattering processes (see, e.g., Derezinski and Gerard, 1997), the roles that Fock space creation and annihilation operators and crossing symmetry play in Weinberg's derivation of CPT invariance have no classical correlates. The algebraic approach, on the other hand, due to its more abstract framework, does seem to have a classical analogue.

Recall from Sections 1.4 and 2.2.1 that Guido and Longo's (1995) algebraic CPT theorem for relativistic quantum field theories makes essential use of the Tomita–Takesaki theorem. This theorem assigns to every pair $(\mathfrak{R}, \Omega)$ of a von Neumann algebra $\mathfrak{R}$ with cyclic vacuum vector Ω, the modular data $(\Delta(t), \mathcal{J})$ consisting of a family of modular operators $\Delta(t)$, $t \in \mathbb{R}$, and a modular conjugation $\mathcal{J}$. The modular operators map $\mathfrak{R}$ to itself, whereas the modular conjugation maps $\mathfrak{R}$ to its commutant $\mathfrak{R}'$. Guido and Longo's proof adopts *modular covariance* (MC), which assumes that the modular operators of the local algebra of a wedge region W implement Lorentz boosts on W. Kuckert's (1995) alternative algebraic proof adopts *modular P_1CT-symmetry*, which assumes that the modular conjugation $\mathcal{J}$ of the local algebra of a wedge W implements a 1-dim PT reflection of it onto its spacelike complement W'. The generalizations of these assumptions that take the forms of Section 2.2.1's *condition of geometric modular action* (CGMA) and *condition of geometric action for the modular groups* (CMG) can similarly be characterized in these terms.

As explained in Section 3.2.4, the Tomita–Takesaki theorem requires separability of the vacuum state, and this fails in NQFTs. This accounts for the failure of algebraic versions of the CPT theorem for NQFTs, but what about algebraic RCFTs? Swanson (2014: 96) points out the following shortcomings. First, since the observables of a classical theory mutually commute, an algebraic formulation of an RCFT should begin with a commutative von Neumann algebra $\mathfrak{R}_C$. The states ω of such an algebra have the property $\omega(AB) = \omega(BA)$, for all $A, B \in \mathfrak{R}_C$ (one says such states are "tracial"). It can now be shown that, for the modular data corresponding to $(\mathfrak{R}_C, \Omega)$, the action of the modular operators $\Delta(t)$ is trivial, i.e., $\Delta^{it} A \Delta^{-it} = A$, $A \in \mathfrak{R}_C$;[25] thus $\Delta(t)$ can no longer be said to implement Lorentz boosts. Moreover, a commutative von Neumann algebra has the property $\mathfrak{R}_C = \mathfrak{R}'_C$, where $\mathfrak{R}'_C$ is the commutant of $\mathfrak{R}_C$. This entails that $\mathcal{J}\mathfrak{R}_C\mathcal{J} = \mathfrak{R}'_C = \mathfrak{R}_C$, hence $\mathcal{J}$ can no longer be said to implement PT reflections. Thus, for algebraic RCFTs encoded in commutative von Neumann algebras, both MC and *modular P_1CT-symmetry* (and their generalizations) loose the bite they need to underwrite derivations of CPT invariance.

At this point, the scoreboard is mixed when it comes to formulating the CPT theorem for RCFTs: purist algebraic RCFTs do not admit such a theorem, and while it can be formulated for pragmatist Lagrangian RCFTs, its physical significance is questionable.

4.3 Summary

Are CPT invariance and the SSC derivable properties in finite NQM and other non-RQFT frameworks? With respect to NQM, if, as is typically the case, particle indistinguishability is cashed out in terms of PI, then particle indistinguishability alone does not entail the symmetrization postulate (the Bose–Fermi alternative). An additional assumption of "wavefunction single-valuedness" is both too narrow and too vague. It is too narrow insofar as it only allows the symmetrization postulate to be recovered for scalar (spin-0) states, and it is too vague in its notion of a multi-valued wavefunction. The latter problem can be addressed by adopting an intrinsic version of the configuration space approach in which a system's wavefunction is defined as a cross-section of an appropriate vector bundle over a multiply-connected configuration space, as opposed to defining it as a function on the (simply-connected) universal covering space of the system's configuration space. This intrinsic version accommodates an understanding of PI as a constraint imposed on a system's algebra of observables, as opposed to a constraint simply imposed on the system's space of states. It also accommodates standard techniques used to quantize topologically non-trivial configuration spaces, hence it side-steps a common complaint of configuration space approaches that they amount to non-standard versions of quantum mechanics. On the other hand, the first problem remains: the symmetrization postulate can only be derived for scalar states. Moreover, the symmetrization postulate is not the SSC! Currently the latter can only be derived for a few highly constrained NQM systems.

Sudarshan's approach, viewed as an attempt to derive the SSC for NQFTs, assumes the fields are Hermitian and possess crossing symmetry, but massive Galilei-invariant quantum fields cannot be Hermitian, and can possess crossing symmetry and the wrong spin–statistics connection (NSSC).

In Chapter 3, we found that CPT invariance was inherently relativistic in the sense that its derivation in RQFTs requires both RLI and a relativistic locality constraint (only the latter in the algebraic approach). In this chapter, we saw that, in principle, CPT invariance is not inherently quantum mechanical. CPT invariance requires both a particular type of spacetime symmetry structure, to underwrite PT invariance, and a decomposition of a theory's state space into matter and antimatter subspaces, to underwrite charge conjugation invariance. The first structure is not available in non-relativistic theories. The second structure can be imposed on both classical and quantum field theories. However, in classical field theories, the map ($C_{\#}$) from matter to antimatter subspaces is not physically relevant. This makes classical CPT symmetry not too interesting, insofar as it is a consequence of imposing a physically questionable map (C) on a more physically relevant symmetry (PT). On the other hand, in quantum field theories, to get a PT symmetry, one needs the physically irrelevant map ($C_{\#}$) to nullify the physically relevant C_* map. This underwrites the intuition that in quantum field theories, C and PT are more connected than they are in classical field theories (as Swanson, 2014: 98 observes). An advantage of the algebraic approach is that it codifies this by making

CPT invariance impossible to derive in RCFTs. On the other hand, an advantage of the Lagrangian approach is that, while it allows the derivation of CPT invariance for RCFTs, it pinpoints where in the derivation physical relevancy can be questioned (i.e., in the form of #-Hermiticity).

NOTES

1. Mickelsson (1984) adopts this approach to derive the SSC for classical point-particle systems in the presence of a background soliton field.
2. One can show that, for non-relativistic theories, the following "first-quantized" approach to encoding statistics is equivalent to the previous "second-quantized" approach (as explained in Baker et al., 2014: 9). Difficulties arise for relativistic interacting theories, however.
3. See, e.g., Earman (2010), Reyes-Lega and Benvides (2010: 1005, 1021), and Landsman (2013).
4. Recall that this means that Q_N contains loops that cannot be continuously deformed into a point. The set of equivalence classes of such loops that cannot be continuously deformed into each other forms a group $\pi_1(Q_N)$ called the first homotopy group of Q_N.
5. See, e.g., Morandi (1992: 116). Berry and Robbins (1997, 2000a, 2000b) make "single-valuedness" of the wavefunction an essential assumption. As various authors point out, a more rigorous approach to the quantization of a classical multiply-connected configuration space defines wavefunctions as sections of an appropriate vector bundle, and the single-valuedness constraint becomes the requirement that such sections be invariant under the action of the structure group of the corresponding principle bundle.
6. See, e.g., Morandi (1992: 128). For $d = 2$, $\pi_1(Q_N) = B_N$, the braid group on N elements. There are many 1-dim representations of B_N, apart from the symmetric and antisymmetric representations. This leads, via equation (4.4) to "anyonic" statistics in two dimensions.
7. In general, this will be the vector bundle associated with a principle bundle with fibers that transform under a given irreducible representation of the group $SU(2)$ that encodes spin. This intrinsic approach is advocated by Bourdeau and Sorkin (1992) and elaborated in Papadopoulos et al. (2004), Benavides and Reyes-Lega (2010), Papadopoulos and Reyes-Lega (2010), Reyes-Lega and Benavides (2010), and Reyes-Lega (2011). Landsman's (2013) analysis of particle indistinguishability is another example, although his goal is to demonstrate the equivalence between approaches to PI that impose it on classical states, and approaches that impose it on quantum observables.
8. See, e.g., Reyes-Lega and Benevides (2010: 1005): "[O]ur main purpose is to draw attention to a point that, in our opinion, appears not to have

been taken sufficiently into account, namely, the relevance of the algebra of operators for any quantum description of a system of identical particles based on the configuration space Q_N. The subtleties involved in the correct definition of infinitesimal (i.e., self-adjoint) generators of symmetries, their domains and the algebras they represent are well known in the context of quantization theory. But, in our view, in the context of the non-relativistic approach to spin–statistics they have received very little attention." The intrinsic approach adopted by Reyes-Lega and Benevides (2010) employs Isham's (1984) canonical group quantization method. Another example is Landsman's (2013) treatment of particle indistinguishability and PI in terms of deformation quantization.

9. Kuckert and Mund (2005) extend this 2-dim result to an arbitrary number of particles.
10. A version of this approach is adopted by Morgan (2004) to derive the SSC for "pseudoclassical" systems that are characterized by non-commuting Grassmann variables.
11. "... we believe that the spin–statistics connection must arise from the properties of the physical system independent of the *processes* that it goes through" (Shaji, 2009: 2–3). "This is acceptable because we do not expect the spin–statistics connection to come out of interactions and other dynamical effects" (Shaji, 2009: 4).
12. Shaji and Sudarshan (2003: 4) and Shaji (2009: 4) focus solely on the first term in equation (4.8a) involving the time derivative of the fields.
13. The demonstration is based on the fact that the antisymmetric metric defined on the carrying space S of representations of $SU(2)$ has the following properties: with respect to this metric, a contraction of integer spin representations is symmetric in the indices of the factors, whereas a contraction of half-integer spin representations is antisymmetric. Step (I)'s symmetry constraint on Λ_{rs} then follows from condition (1) that the Lagrangian density (4.8) is an $SU(2)$ scalar, which means that it must be built out of contractions of $SU(2)$ fields under this scalar product.
14. To demonstrate this, one assumes that the surface variation of the action $I = \int d^4x\mathcal{L}_{kin}(x)$ is the generator of infinitesimal transformations of the fields on the boundary surfaces. This is expressed by the condition $[\xi_r, \delta I] = \delta\xi_r$. By assuming either commutation or anticommutation relations for the fields in this expression (i.e., by assuming the Bose–Fermi alternative), one obtains step (II)'s symmetry constraint on Λ_{rs}.
15. In contrast, recall from Section 3.2.1 that a Poincaré-invariant quantum field ϕ_j transforms under a unitary representation $U(g)$ of the universal covering $\tilde{P}_+^\uparrow$ of the restricted Poincaré group according to $U(g)\phi_j(x)U(g)^{-1} = \sum S_{jk}(A^{-1})\phi_k(x')$, where $g \in \tilde{P}_+^\uparrow$, $x' = \Lambda x + a$, and $S_{jk}(A^{-1})$ is a unitary finite dimensional matrix representation of $A \in SL(2, \mathbb{C})$. The absence of a phase factor entails that the Hermitian conjugated field transforms in exactly the same way.

16. An analysis of Bell's theorem is given in Greaves (2008, 2010).
17. The theorem, as reported by Greaves (2008: 58) states that if $f : \mathbb{C} \to \mathbb{C}$ is an everywhere complex-differentiable function of a single complex variable which vanishes on the real line, then it vanishes on all of $\mathbb{C}$.
18. The remainder of the proof runs as follows: under the PT transformation $\Lambda_{\mu\nu} = -\delta_{\mu\nu}$, expression (4.11) becomes $\mathcal{F}[(-1)^{n(T)} T_{\mu\nu\ldots}(x)] = (-1)^n \mathcal{F}\left[T_{\mu\nu\ldots}(x)\right]$, where on the left-hand side, $n(T)$ is the rank of $T_{\mu\nu\ldots}$, corresponding to the number of $-\delta_{\mu\nu}$s (one per index of $T_{\mu\nu\ldots}$), and on the right-hand side, n is the rank of $G_{\alpha\beta\ldots}$. Since this is supposed to hold at all spacetime points, it holds in particular at $-x$. The transformation rule for $\mathcal{F}$ under PT then becomes $\mathcal{F}\left[T'_{\mu\nu\ldots}(x)\right] = (-1)^n \mathcal{F}\left[T_{\mu\nu\ldots}(-x)\right]$, where $T'_{\mu\nu\ldots}(x) = (-1)^{n(T)} T_{\mu\nu\ldots}(-x)$. This indicates that $\mathcal{F}$ is PT invariant.
19. A complex structure on V is a map $\mathcal{J} : V \to V$ such that $\mathcal{J}^2 = -1$. $\mathcal{J}$ defines a rule for multiplying elements of V with complex numbers, $(a + ib)\, v = av + b\mathcal{J}v$, for $a, b \in \mathbb{C}$, $v \in V$. A symmetry of a Lagrangian field theory is a transformation on fields that leaves the action $S = \int \mathcal{L} d^4x$ invariant.
20. The action of $\mathcal{G}$ may be irreducible on V, in which case the quantized theory will not admit particle and antiparticle subspaces. The necessary and sufficient condition for reducibility is the presence of a complex structure $\mathcal{J}$ on V that commutes with $\mathcal{G}$.
21. The difference between classical charge conjugation C and Q-conjugation mirrors the difference between the two distinct types of complex structure associated with Lagrangian QFTs obtained from field quantization, namely, the complex structure associated with complex classical fields, and the complex structure associated with the single-particle Hilbert space.
22. This follows from the fact that C anticommutes with $\mathcal{J}$, and particle and antiparticle states are eigenstates of $\mathcal{J}$ with eigenvalues +1 and −1 respectively. Thus $\mathcal{J}(C|\psi\rangle) = -C\mathcal{J}|\psi\rangle = -C(\pm|\psi\rangle) = \mp C|\psi\rangle$. Hence $C|\psi\rangle$ is an antiparticle (resp. particle) state if and only if $|\psi\rangle$ is a particle (resp. antiparticle) state.
23. Normal-ordering is sufficient to ensure the well-definedness of products of fields that can be decomposed into positive and negative frequency modes. Most realistic interacting fields cannot be so decomposed.
24. The so-extracted theorem is a generalization of Bell's theorem that extends it to half-integer spin representations of $L_+^\uparrow$.
25. More precisely, the modular automorphism group $\{\sigma_t\}$ induced by the action of the modular operators via $\sigma_t(A) = \Delta^{it} A \Delta^{-it}$, $A \in \mathfrak{R}$, is trivial if and only if the vacuum state ω corresponding to Ω is tracial (Summers, 2006: 252).

5

What Explains CPT Invariance and the Spin–Statistics Connection?

This chapter addresses the question, "What explains CPT invariance and the spin–statistics connection (SSC)?" To set the stage, Section 5.1 reviews the reasons why we should care about explaining CPT invariance and the SSC. Section 5.2 then considers the extent to which the CPT and spin–statistics theorems in relativistic quantum field theories (RQFTs) can be said to provide us with the sought after explanations. Based on the discussion in Chapters 1–4, I will argue that these theorems cannot be understood as providing any of the standard types of explanation that have appeared in the philosophy of science literature: deductive–nomological, unifying, causal, or structural. They do not provide deductive–nomological explanations to the extent that they cannot be viewed as unique derivations from first principles. They do not provide unifying explanations to the extent that they cannot be understood as a part of a unifying argument pattern endorsed by a consensus of the RQFT community. They do not provide causal explanations to the extent that causal explanations can be understood as appealing to constraints on a physical system's space of *dynamically* possible states, and the CPT and spin–statistics theorems place constraints on a physical system's space of *kinematically* possible states. Finally, they cannot be said to provide structural explanations of CPT invariance and SSC, insofar as the essential structural aspects of the theorems differ from approach to approach, these structural aspects hold only for RQFTs and hence cannot explain the essential role that SSC in particular plays in non-relativistic theories, and these structural aspects cannot explain why *realistic interacting* RQFTs possess CPT invariance and SSC. I shall argue that these problems with structural explanations make trouble for the "geometric" explanations of CPT invariance offered by Greaves (2010) and Swanson (2014).

These considerations suggest that a full account of CPT invariance and SSC must go beyond the CPT and spin–statistics theorems in RQFTs. A full account should explain by virtue of both a derivation from non-fundamental principles in RQFTs, and an explication of intertheoretic relations between RQFTs on the one hand, and non-relativistic quantum field theories (NQFTs) and non-relativistic quantum mechanics (NQM) on the other. Section 5.3 characterizes this type of explanation and compares it with a similar account given by Weatherall (2011). In Weatherall's example, a general observational feature of the world (the equality of the gravitational and inertial mass of any body) that is expressed in one theory, Newtonian gravity, is explained by appealing to another, presumably more fundamental theory, general relativity (GR), in which the *explanandum* (the thing to be explained) cannot be expressed. The explanatory work is done by means of a translation between GR and Newtonian gravity which demonstrates how the explanandum arises in a limiting process that goes from GR to Newtonian gravity. In this chapter, a similar explanation will be constructed for SSC in interacting NQFTs and NQM. This explanation demonstrates how SSC arises in a limiting process that goes from realistic interacting RQFTs (in which SSC cannot be expressed), to non-interacting RQFTs (in which SSC can be expressed and derived), and thence to interacting NQFTs and NQM (in which SSC appears as a brute empirical fact). On the other hand, I argue that the presence of both SSC and CPT invariance in realistic interacting RQFTs cannot be explained by even this type of explanation, and that currently, the *existence problem* precludes a full understanding of these properties in realistic interacting RQFTs.

5.1 Why CPT Invariance and the Spin–Statistics Connection Need Explanations

Why should we care about explaining CPT invariance and SSC? Let's first consider CPT invariance. Recall that it is an essential characteristic of RQFTs, and hence, presumably, an essential property of the basic constituents of the world. It links an internal gauge symmetry, embodied in the operation of charge conjugation, with spacetime symmetries, embodied in the operations of spatial parity and time reversal. In doing so, it underwrites an interpretation of matter which assigns to every matter state, an antimatter state. It entails not just the existence of antimatter, but also that antimatter partners to ordinary matter have equal masses and lifetimes, equal but opposite spins and magnetic moments (if any), and opposite charges. These predictions have been confirmed by experiments that Duncan (2012: 483) describes as "... among the most precisely tested predictions of quantum field theory." Equality of masses, for instance, has been tested in high energy experiments involving neutral kaon decays. Based on data on CP violation in such processes, the limit on the mass difference between a neutral kaon K^0 and its antiparticle $\bar{K}^0$ is given by (Olive et al., 2014: 96)

$$|(m_{\bar{K}^0} - m_{K^0})/m_{K^0}| \leq 0.6 \times 10^{-18}.$$

In addition to the empirical evidence for CPT invariance, one can argue that it plays a fundamental explanatory role in the reconciliation, in RQFTs, of relativity and quantum mechanics. Historically, the incorporation of relativity into quantum mechanics required the balancing of three principles: Lorentz invariance, positivity of energy, and locality. Imposing the first on the Schrödinger equation results in a relativistic equation with positive energy solutions that violate locality.[1] This suggests a move from the parabolic single-particle Schrödinger wave equation to hyperbolic single-particle wave equations (i.e., the Klein–Gordon and Dirac equations). This move upholds relativity and solves the locality problem, but at the expense of positivity of energy: while solutions to hyperbolic partial differential equations have compact support, they contain both positive and negative frequency components, the latter violating positivity of energy.[2] CPT invariance can be understood as a solution to this balancing act (Swanson, 2014: 62–63): it allows the negative frequency solutions of hyperbolic wave equations that describe *single-particle* states to be reinterpreted as positive energy solutions that describe *multi-particle* antimatter states with the same mass, spin and lifetime, but opposite charge, as positive energy multi-particle matter solutions. Thus, according to Swanson (2014: 63), CPT invariance "... is therefore a result with profound foundational significance. A breakdown of PCT [CPT] symmetry would signal the need for a radical overhaul of our understanding of relativistic quantum theory."

With respect to SSC, physicists point out that it has a "profound impact" (Zee, 2010: 121) in NQM and NQFT:

> From the microscopic structure of atoms to the macroscopic structure of neutron stars, a dazzling wealth of physical phenomena would be incomprehensible without this spin–statistics rule. Many elements of condensed matter physics, for instance, band structure, Fermi liquid theory, superfluidity, superconductivity, quantum Hall effect, and so on and so forth, are consequences of this rule. (Zee, 2010: 120)

> The world would be a different place if spin-one-half particles were not subject to Pauli's exclusion principle. In all fundamental branches of modern (natural) science, the connection between particle spins and multi-particle behavior plays a crucial role, and to date, no physical system violating it has ever been observed. (Kuckert, 2007: 207)

Recall from Section 1.1.2 that Fermi–Dirac statistics requires fermionic multi-particle states to be antisymmetric under a permutation of single-particle substates, and this entails that multi-particle states cannot consist of single-particle substates that agree on all their non-spatiotemporal properties. Thus Fermi–Dirac statistics entails an exclusion principle: any two fermions cannot be in the same single-particle substate. I will refer to this principle as the Pauli exclusion principle (PEP), with the proviso that Pauli's original formulation was restricted to electrons in atoms. Bose–Einstein statistics, on the other hand, allows bosonic multi-particle states to consist of multiple copies of the same single-particle substate, i.e., it allows a collection of bosons to all occupy the same single-particle state. According to the standard model, fundamental matter states (leptons and quarks) possess half-integer spins, whereas fundamental gauge field states (carriers of the electromagnetic, strong and weak forces) possess integer spins. Thus SSC, in conjunction with the standard model, entails that fundamental matter states must be fermions that obey PEP. In particular, since electrons are spin-1/2 fermions, the orbitals they enter into in atoms are filled in such a way that no two electrons can occupy the same orbital state. This provides the basis for an explanation of the structure of the periodic table, and hence the stability of matter. In this and similar examples, SSC figures into explanations of non-relativistic interacting quantum systems best described by finite-dimensional NQM.

SSC also figures into explanations of non-relativistic interacting physical systems best described as possessing infinite degrees of freedom, and hence represented by interacting NQFTs. Examples include condensed matter systems that exhibit collective behavior. For instance, a Bose–Einstein condensate (BEC) is characterized by a phase transition at which some proportion of a collection of non-interacting bosons forms a condensate by all coming to occupy the same zero energy state (see, e.g., Annett, 2004: 6). The mechanism of BEC formation underwrites explanations of many other condensed matter systems. For instance, there are only two superfluids which can be studied in the laboratory: helium-4 and helium-3 (Annett, 2004: 21). Both have identical electronic properties, but SSC entails that a helium-4 atom is a boson (consisting of an even number of spin-1/2 nucleons), while a helium-3 atom is a fermion (consisting of an odd number of spin-1/2 nucleons). This subsequently entails they have different properties at low temperatures. The superfluid properties of helium-4 are derived by describing its formation in terms of a BEC comprised of helium-4 atoms. The superfluid properties of helium-3 are derived by describing its formation in terms of a BEC comprised of "composite bosons" that consist of pairs of helium-3 atoms. Among other things, this allows for three distinct helium-3 superfluid phases. The process whereby composite bosons made up of pairs of fermions condense to form a BEC also underwrites the standard Bardeen–Cooper–Schrieffer (BCS) theory of superconductivity, in which the composite bosons are pairs of interacting electrons ("Cooper pairs"). Theories of unconventional superconductors also make use of this explanatory device, the only difference being that the electron pairs obey different symmetries than conventional Cooper pairs. A final example of a condensed matter system that makes use of the BEC explanatory framework is a fractional quantum Hall liquid. This system arises when electrons moving in a two-dimensional conductor are subjected to a strong external magnetic field. For particular values of the external field, the system acquires properties characteristic of superconductors (incompressibility and zero resistivity). One can show that at these values, a description of the interacting electron system is equivalent to a description of a system of bosons coupled to both an external magnetic field and an internal Chern–Simons gauge field (Zhang, 1992). The Chern–Simons interaction can be characterized as coupling an odd number of internal magnetic fluxes to each boson. In this case, the resulting object (still referred to as a composite boson) mimics an original electron in the sense of obeying Fermi–Dirac statistics.[3] The strength of the Chern–Simons coupling is chosen so that, at the critical BEC temperature, it cancels the coupling to the external magnetic field, and the system behaves as if it was comprised of non-interacting bosons, which subsequently condense to form a BEC.[4]

These and other examples underwrite the "profound impact" that SSC has in interacting NQFTs and NQM. Now the same physicists who point out this impact of SSC in *non-relativistic* quantum theories also claim that an explanation of SSC had to wait until the advent of *relativistic* quantum field theories:

> ... the explanation of the spin–statistics connection by Fierz and by Pauli in the late 1930s, and by Luders and Zumino and by Burgoyne in the late 1950s, ranks as one of the great triumphs of relativistic quantum field theory. (Zee, 2010: 121)
>
> [The spin–statistics theorem] ... clarifies one of the great mysteries of non-relativistic quantum theory: the contrasting symmetry properties of the wavefunctions of particles of integer (boson) versus half-integer (fermionic) spin. (Duncan, 2012: 59)

But not all authors are satisfied with this appeal to the spin–statistics theorem as an explanation of SSC:

> The spin–statistics connection seems crucial to understanding the behavior of several physical systems for which relativistic considerations seem quite insignificant... Non-relativistic theories seem to adequately describe most of these systems and the spin–statistics connection has to be inserted 'by hand' when formulating these theories. (Shaji, 2009: 2)
>
> An explanation has been worked out by Pauli from complicated arguments of quantum field theory and relativity... we have not been able to find a way of reproducing his arguments on an elementary level... This probably means we do not have a complete understanding of the fundamental principle involved. (Feynman, 1965: 4–3)

The concerns raised by Shaji and (famously) Feynman are concerns over the extent to which the spin–statistics theorem in RQFTs explains the presence of SSC in non-relativistic theories. This is significant, since the overwhelming majority of evidence for SSC comes from observations of physical systems best described by NQFTs and NQM. This is to be contrasted with CPT invariance, for which the evidence comes from physical systems best described by RQFTs. Thus we should distinguish between the following two questions:

(I) Do the CPT and spin–statistics theorems explain CPT invariance and SSC in RQFTs?

(II) What explains SSC in non-relativistic theories?

The implicit assumption in Shaji and Feynman, and arguably, the received view among physicists, is that the answer to question (I) is "yes" and that there is currently no feasible answer to question (II). Sections 5.2 and 5.3 will argue against these positions. The goal of Section 5.2 is to argue that the answer to question (I) is "no:" the CPT and spin–statistics theorems, by themselves, do not explain CPT invariance and SSC in RQFTs, at least under contemporary notions of scientific explanation. The goal of Section 5.3 is to argue that, Section 5.2 notwithstanding, an adequate explanation of SSC in non-relativistic theories can be constructed.

5.2 The CPT and Spin–Statistics Theorems Do Not Explain CPT Invariance and the Spin–Statistics Connection

Chapters 1–4 told us that the CPT and spin–statistics theorems in RQFTs are characterized by the following features:

(a) Proofs of the theorems can be formulated in mathematically and conceptually distinct ways.
(b) The theorems do not hold for realistic interacting RQFTs.
(c) The theorems do not hold for NQM and NQFTs.

Claim (a) is based on the discussion in Chapter 1 of the versions of the CPT and spin–statistics theorems in the Wightman axiomatic approach to RQFTs, Weinberg's approach, the Lagrangian approach, and the algebraic approach. As explained in Section 1.3, these different versions of the theorems are conceptually distinct: not only do they differ on the principles needed to derive CPT invariance and SSC, they also differ on what they take to be the objects that possess these properties (fields, particles, observables). Moreover, as explained in Section 1.4, they differ over how they treat interactions. In particular, purist approaches (Wightman axiomatic and algebraic) adopt a more rigid definition of existence for interacting RQFTs, while pragmatist approaches (Weinberg and Lagrangian) allow (implicitly) for weaker notions. The fact that none of these notions of existence holds for the realistic interacting RQFTs associated with the empirically successful standard model (which serves as the basis for the observations that confirm CPT invariance) was Section 1.4's *existence problem*, which underwrites claim (b). Finally, the discussion in Chapters 2–4 underwrites claim (c).

In the rest of this section, I will argue that these results make it problematic to claim that the CPT and spin–statistics theorems provide explanations of CPT invariance and SSC in RQFTs, at least under standard accounts of explanation. To set the stage, I would like to be clear on what I take to be the *explanandum*, i.e., the thing to be explained. I take CPT invariance and SSC to be properties that a physical system may possess. As a matter of fact, it is observed that all physical systems possess these properties. The question then arises, Why is this the case? The typical response in the physics literature adopts the following explanation:

> The states of physical systems described by RQFTs are characterized by a set of *fundamental* properties, and CPT invariance and SSC can be derived from this set insofar as, if the state of a physical system possesses these fundamental properties, then it must possess CPT invariance and SSC. (∗)

In this explanation, the explanandum is a general regularity, as opposed to a particular fact. We wish to know why *all* physical systems (of the relevant type) possess CPT invariance and SSC, as opposed to why any particular physical system possesses them. Claim (∗) purports to explain this regularity by deriving it from a set of other regularities (alternatively, laws or principles) of the form "All physical systems possess fundamental property X," where X can be *restricted Lorentz invariance* (RLI), the *spectrum condition* (SC), *modular covariance* (MC), etc. (i.e., one of the principles associated with any of the approaches to the CPT and spin–statistics theorems reviewed in Section 1.2). The task of the rest of this section is to compare claim (∗) with the standard accounts of explanation; namely, deductive–nomological, unificationist, causal, and structural.

5.2.1 The Deductive–Nomological Account

A deductive–nomological (DN) explanation explains by virtue of a derivation from one or more covering laws together with a specification of antecedent conditions required to apply the laws to the given explanandum.[5] DN explanations demonstrate how the explanandum is nomically expected, i.e., how it follows necessarily from lawlike premises. There are at least two interrelated concerns with viewing the CPT and spin–statistics theorems as providing DN explanations of CPT invariance and SSC, both stemming from claim (a), the fact that the theorems admit conceptually distinct formulations. One of these concerns has to do with the nature of the covering laws, in this case, the principles, such as RLI, *cluster decomposition* (CD), MC, etc., that underwrite the various versions of the theorems. The other concern has to do with the nature of the explanandum; in this case, the general regularity that all physical systems of the relevant type possess CPT invariance and SSC.

An initial concern is that the notion of a covering law in a DN explanation is a matter of debate among philosophers of science. Among other things, laws have been construed as contingent or necessary relations among universals, as general regularities, or as regularities that underwrite the best (simplest, strongest, etc.) systematization of the facts. To be charitable, I will assume that the various principles (RLI, the SC, CD, MC, etc.) that underwrite alternative proofs of the CPT and spin–statistics theorems can be viewed as laws, however one chooses to describe the latter. The more pressing concern is the extent to which these principles can be considered fundamental.

The issue of fundamentality arises in the context of the second concern with viewing the CPT and spin–statistics theorems as DN explanations of CPT invariance and SSC. Again, I take the latter to be associated with general regularities (alternatively, laws or principles), i.e., taken separately, these regularities are the observed facts that all physical systems of the relevant type possess CPT invariance and SSC. The concern is that, notoriously, the DN account has problems in explaining general regularities. Hempel and Oppenheim's (1948: 159) original account restricted the explanandum of a DN explanation to a particular fact due to the following problem (call it the problem of conjunctions): a law L_1 can be derived from the conjunction L_1 & L_2 of that law with any other (true, empirical) law L_2. This conjunction may itself be considered a law and hence can appear as a premise of a DN explanation of L_1. But the derivation of L_1 from L_1 & L_2 should not necessarily be taken to explain L_1 (take L_1 to be Kepler's laws and L_2 to be Boyle's Law). As Psillos (2007: 135) notes, the problem of determining when a derivation of a law from other laws counts as a legitimate explanation of the former is the issue of what makes one law more fundamental than another: a derivation of L_1 from L should count as an explanation of L_1 whenever L counts as more fundamental than L_1 (thus the conjunction of Kepler's law and Boyle's law doesn't explain Kepler's law since, intuitively, the former is not more fundamental than the latter). One attempt at making this distinction, proposed initially by Friedman (1974), identifies a fundamental law with a unifying law. The next section will consider Kitcher's (1989) version of unificationism, which constitutes a distinct approach to explanation than DN. At this point, we need only understand the motivation for this approach; namely, that a law is unifying, and thus fundamental, if it belongs to a small set of laws from which a large body of claims can be derived. Under one gloss of this intuition, a principle like CPT invariance or SSC is explained just when it can be uniquely derived from a set of fundamental (viz., unifying) principles.

The upshot of this discussion so far is that a DN explanation cannot explain CPT invariance and SSC without facing the problem of conjunctions, and under one proposed solution to this problem, a principle is explained by a unique derivation of it from a set of more fundamental principles. The problem with trying to understand CPT invariance and SSC in this context is that there is no unique set of first principles from which these properties can be derived in RQFTs. The existence of conceptually distinct alternative formulations of these theorems indicates there is no unique derivation of the properties; and it also puts into question whether the principles used to derive these properties can be considered fundamental. Intuitively, if CPT invariance and SSC are to be explained by demonstrating how they are nomically expected from fundamental first principles, there should be a unique nomic story to tell about each of them. The CPT and spin–statistics theorems do not provide us with such stories.

Moreover, this DN attempt at understanding CPT invariance and SSC assumes that, within a given approach to RQFTs, CPT invariance and SSC are less fundamental than the principles employed to derive them. Weatherall (2014) has suggested a view of theories that questions whether the central principles associated with a theory can be ordered in terms of fundamentality. According to this view, "Instead of thinking of the foundations of a physical theory as consisting of a collection of essentially independent postulates from which the rest of the theory is derived, one might instead think of the foundations of a theory as consisting of a network of mutually interdependent principles—a collection of interlocking pieces..." (Weatherall, 2014: 26). Weatherall's example concerns the derivation of the geodesic principle in GR and geometrized Newtonian gravity. In both cases, this derivation requires a conservation principle and assumptions about energy conditions, and there are reasons to believe that none of these principles is more fundamental than the others. Indeed, Weatherall's "puzzleball" view of theories suggests that "... the *central principles* of some scientific theories are mutually interderivable" (Weatherall, 2014: 33). Naively, one might take the existence of conceptually distinct approaches to RQFTs as additional evidence for the puzzleball view. These approaches indicate that an RQFT can be formulated in terms of different sets of interlocking principles.[6]

5.2.2 The Unificationist Account

Do the CPT and spin–statistics theorems provide unifying explanations of CPT invariance and SSC? Under Kitcher's (1989) account, a unifying explanation explains by virtue of belonging to the most unifying systematization of the set K of claims currently endorsed by the scientific community. A systematization Σ of K consists of a subset of statements in K from which the rest of K can be derived. Σ is unifying if it maximizes scope, simplicity, and stringency.[7] The most unifying systematization is called the explanatory store $E(K)$ over K. Thus to determine if the CPT and spin–statistics theorems provide unifying explanations of CPT invariance and SSC, we need to determine if they belong to $E(K)$ for the relevant K.

Kitcher (1989: 431) assumes K is consistent and deductively closed, and $E(K)$ is unique.[8] This presents problems for viewing the CPT and spin–statistics theorems as members of $E(K)$. Note first that we can restrict K to those claims associated with RQFTs; but there are multiple, conceptually distinct formulations of RQFTs. We can either restrict K to claims associated with just one of these approaches, or we can include in K claims associated with all approaches. Note that if we do the latter, we will include in K claims associated with both pragmatist and purist approaches to RQFTs. But then K will not be consistent. It will contain statements like "QED (as a model of the relevant purist axioms) does not exist," and "QED (as a pragmatist truncated perturbative expansion in the relevant interaction Hamiltonian) does exist." Perhaps less controversially, it will contain statements like "CPT invariance entails SSC" (as Section 1.3 observed for pragmatist versions of the CPT and spin–statistics theorems), and "CPT invariance does not entail SSC" (for purist versions). Thus if K is to be consistent, we will have to restrict it to the claims associated with just one approach to RQFTs, or at least one family of approaches, either pragmatist or purist. Suppose, then that K contains only those claims associated with pragmatist approaches. Then assumedly $E(K)$ contains pragmatist versions of the CPT and spin–statistics theorems, but not purist versions. Intuitively, purist versions will not even be in K. Moreover, if we put them in $E(K)$ by hand, we will make the latter larger without increasing its scope, thus decreasing its simplicity, and it will no longer be the most unifying systematization of K. Similarly, if we assume K contains only those claims associated with purist approaches to RQFTs, then pragmatist versions of the CPT and spin–statistics theorems will not be in $E(K)$. This suggests that, in order to view the CPT and spin–statistics theorems as providing unifying explanations of CPT invariance and SSC, we will have to first choose between which approach to RQFTs to adopt, pragmatist or purist. This seems a bit too constraining. Indeed in practice, such a choice is typically not made. On the one hand, the practice among pragmatists suggests a combination of pragmatism and purity: we are typically told immediately after being presented with pragmatist versions of the CPT and spin–statistics theorems to consult purist versions for further details. On the other hand, purists certainly do not intend to reject all the claims associated with pragmatist approaches (particular claims about the values of scattering amplitudes, say, and in general, all the empirical claims associated with high energy particle physics that are derived using pragmatist techniques).

In short, if we assume that the set K of claims associated with RQFTs is both consistent and deductively closed, then the best systematization $E(K)$ of K will not be unique: there will minimally be purist and pragmatist versions. On the other hand, if we allow K to encompass all claims associated with RQFTs, both pragmatist and purist, then it will not be consistent. Note that we might expect this type of situation to arise in areas of immature science in which the body of claims has not yet been consistently systematized. In particular, it might be expected that unifying explanations of the phenomena associated with an immature science cannot be given. On the other hand, irrespective of how a distinction between an immature and a mature science is made, one would expect that RQFTs should count as prime examples of the latter, given their immense empirical success.

Thus, in general, since there is no consensus on how to formulate RQFTs, let alone on how to formulate the CPT and spin–statistics theorems, the claim that they offer unifying explanations of CPT invariance and SSC is suspect.

5.2.3 Causal Accounts

A causal explanation explains by virtue of specifying a cause. On the surface, it's not immediately clear how an explanation of CPT invariance and SSC based on an appeal to the CPT and spin–statistics theorems can be interpreted as specifying causes. Note first that the intended explananda are regularities, and most discussions of causal explanation are restricted to explanations of particular events. There are, however, exceptions. Strevens (2008: 220), for instance, identifies two types of causal explanation of regularities: a metaphysical causal explanation of a regularity identifies a "rich and suitably objective relation of metaphysical dependence" between the regularity and more fundamental laws, whereas a mechanistic causal explanation of a regularity identifies a mechanism responsible for the regularity.[9] On the surface, the CPT and spin–statistics theorems make no explicit reference to relations of metaphysical dependence, nor do they refer to mechanisms that might underwrite CPT invariance and SSC.[10]

Another causal account of regularity explanation can be found in Woodward (2003). Woodward considers an adequate explanation to provide both causal and counterfactual information about its explanandum; the latter insofar as an explanation should exhibit a pattern of counterfactual dependence between explanans and explanandum. Woodward (2003: 187) cites, as an example, an explanation for the regularity encoded in the general expression for an electric field due to a charged line source, $E = (1/2\pi\varepsilon_0)(\lambda/r)$, where λ is the charge per unit length. This expression can be derived from Coulomb's law in conjunction with relevant boundary conditions. According to Woodward, the derivation provides both counterfactual and causal information. Counterfactually, it shows how the expression would differ, depending on different boundary conditions. Causally, it grounds the possibilities of differing boundary conditions in causes, in the sense that the expression "... conveys information that is relevant to manipulating or controlling the field" (Woodward, 2003: 196). At first glance, an explanation of CPT invariance and SSC by an appeal to the CPT and spin–statistics theorems might be thought of as another example. Taken literally these theorems demonstrate how the imposition of a set of constraints on the space of physically possible states of an RQFT entails that such states must possess CPT invariance and SSC. The proofs of the theorems demonstrate how CPT invariance and SSC depend counterfactually on the constraints, i.e., they show how, if one or more of the constraints are not imposed, then the possible states of an RQFT would not possess CPT invariance and SSC. In this sense, they convey counterfactual information about these properties. But in what sense, if any, do they convey causal information?

Taken literally, the CPT and spin–statistics theorems entail that, if the state of a physical system is constrained in certain ways (i.e., if it is characterized by RLI, the SC, etc.), then it must be constrained in an additional way (i.e., it must be characterized by CPT invariance or SSC). These constraints (in addition to others) then act to restrict the possible ways the state can evolve in time by means of dynamical equations of motion. A dynamical trajectory can be thought of as a path in a state space that connects an initial state with a final state by means of a dynamical map that encodes an equation of motion. This map defines what might be called a dynamical entailment relation between states. If dynamical entailment supervenes on causal dependence, i.e., if whenever two states are dynamically related, they are causally related, then a dynamical trajectory supervenes on what Lewis (1986: 215) calls a causal history.[11] Thus, under a rather charitable understanding of dynamics and causes (under which we associate a dynamical trajectory with a causal history), it might be suggested that the CPT and spin–statistics theorems explain CPT invariance and SSC by virtue of providing information about the possible causal histories of RQFT states.[12]

Apart from the amount of charity needed to associate dynamical trajectories with causal histories, this suggestion faces a problem with how the constraints associated with CPT invariance and SSC are typically understood. In all the approaches to RQFTs discussed in Chapter 1, there is an implicit distinction between two types of constraints imposed on a theory's state space, *kinematical* constraints and *dynamical* constraints. Both types jointly determine the space of physically possible states of the theory, but do so in different ways. One first identifies a space of kinematically possible states on the basis of symmetry principles and definitions of the types of physical systems one is describing (recall in the context of classical field theories that this involves, among other things, a map from a spacetime manifold to a space of field values, where the latter may have additional types of structures defined on it; in the algebraic approach, the kinematics is specified, in part, by the algebra of observables and various axioms). One then imposes a dynamics on the space of kinematically possible states. Again, this amounts to a map that takes initial states to final states, and defines the space of dynamically possible states as a subset of the space of kinematically possible states. In all the approaches to the CPT and Spin–Statistics theorems, the way the split between kinematics and dynamics is performed places symmetry principles (like CPT invariance) and superselection sectors (like those defined by statistics) on the side of kinematics, i.e., as constraints imposed on a theory's state space prior to the specification of the theory's dynamics.

Thus, in all the standard formulations, the CPT and spin–statistics theorems place constraints on *kinematically*, as opposed to dynamically, possible states. Both theorems say that any state of a physical system described by an RQFT must possess CPT invariance and SSC, *regardless* of what dynamics it satisfies (i.e., regardless of the theory's equations of motion). Thus even a charitable (and admittedly naive) association between dynamical trajectories and causal histories cannot underwrite an understanding of the CPT and spin–statistics theorems as providing causal information about CPT invariance and SSC.[13] Note that kinematical constraints do provide some information about dynamical trajectories, and hence (naively) causal histories, insofar as the dynamically possible states of a theory are a subset of its kinematically possible states. But, arguably, this information is causally irrelevant insofar as it is robust under variations of the dynamics.[14] Thus the kinematical information (of this nature) associated with a theory doesn't tell you what would happen if the theory's dynamics had been slightly different.

A possible objection to the argument so far might run as follows: the distinction between kinematics and dynamics for a given theory is not absolute, but rather a matter of convention. Thus it should always be possible, in principle, to reinterpret a kinematical constraint as a dynamical constraint. This suggests that CPT invariance and SSC can be interpreted as dynamical constraints that restrict how the theory's possible states evolve in time. And this suggests, under a charitable understanding of the relation between dynamics and causes, that CPT invariance and SSC are causal in nature. Spekkens (2013: 2) argues for such a conventionality of kinematics. He claims that how one chooses to distinguish kinematics from dynamics makes no empirical difference; rather both kinematics and dynamics supervene on causal structure, and it is the latter in which explanatory power resides. As an example, Spekkens cites the Newtonian and Hamiltonian formulations of classical mechanics. These formulations posit distinct spaces of kinematically possible states (configuration space versus phase space), and distinct dynamics (second-order Euler–Lagrange equations versus first-order Hamilton equations), but agree on all empirical predictions. Thus:

> It's not possible to determine which kinematics, Newtonian or Hamiltonian, is the *correct* kinematics. Nor can we determine the correct dynamics in isolation. The kinematics and dynamics of a theory can only ever be subjected to experimental trial as a pair. (Spekkens, 2013: 2)

In principle then, CPT invariance and SSC, and the constraints that entail them, can be interpreted as dynamical constraints, as long as we adjust the kinematical aspects of the theories they appear in appropriately. In fact, some authors have suggested an interpretation of the correlations associated with Bose–Einstein and Fermi–Dirac statistics as effective forces of attraction and repulsion, respectively. In particular, Bose–Einstein statistics allows a collection of bosons to all be in the same state, and this suggests an effective force of attraction. Fermi–Dirac statistics, on the other hand, entails the PEP which prohibits fermions from occupying the same state, and this suggests an effective force of repulsion.[15] However, there are good reasons to reject this "real force" interpretation of quantum statistics as causal in nature; or at least as providing the sort of information one should expect of a causal explanation; namely, that of providing counterfactual information. Mullin and Blaylock (2003: 1228) observe that the forces identified by the real force interpretation are supposed to be effects of the spatial part of a multiparticle system's wavefunction, and if the system possesses spin, its statistics must be encoded in a total wavefunction that combines spatial and spin degrees of freedom. Thus a total wavefunction may be symmetric even though its spatial part is antisymmetric, so long as its spin component is also antisymmetric. This may lead to examples of fermionic attractive forces and bosonic repulsive forces. Thus an explanation of PEP in terms of a repulsive force doesn't seem to convey counterfactual information of the relevant sort: if a state were not fermionic, it is not necessarily the case that it would not experience an effective repulsive force (i.e., under the real force interpretation, it is possible for bosonic states to experience an effective repulsive force).[16]

In general, one might agree that the kinematical/dynamical distinction is conventional, but still maintain that some constraints have a particular invariant status, insofar as they are robust (i.e., remain unchanged) under variations of the dynamics. Thus whether one chooses to call the constraints imposed by the CPT and spin–statistics theorems kinematical or dynamical may be a matter of convention, but in all theories in which they appear, they remain unaffected under changes in the dynamics. To the extent that such dynamical invariants do not track changes in dynamics, and hence (perhaps naively) changes in causal structure, such invariants seem independent of the latter.

The concept of a dynamically robust constraint can provide some clarification of Skow's (2014) analysis of the relation between constraints and dynamics.[17] Skow (2014: 462) considers three possible ways a constraint law (i.e., a law that imposes a constraint on a space of states) may be related to dynamical laws:

(a) The dynamical laws are suspended in favor of the constraint law at appropriate times.
(b) The constraint law is a derivative law that follows from (and is explained by) the dynamical laws.
(c) The constraint law is more fundamental than the dynamical laws, and thus explains why the dynamical laws are constrained in the way they are.

Skow considers, as an example, the PEP, viewed as a constraint on the state space of a theory in NQM. In this example, the dynamical laws are encoded in the Schrödinger equation, and Skow points out that option (a) cannot be the case: unitary Schrödinger dynamics leaves superselection sectors that encode statistics invariant, thus the dynamical laws in this case are never suspended; rather, they are constrained (by unitarity and by SSC) to act within and only within subsets of possible states defined by superselection sectors. On the other hand, Skow considers the choice between options (b) and (c) to be up in the air.[18] But option (b) is ruled out, in this particular case, by the discussion in the preceding paragraph: SSC, and hence the PEP, is robust under changes in the dynamical laws; hence, it cannot be considered derivative of the latter. Option (c) is left, and indeed it seems to be the standard way that NQM and RQFTs encode kinematical constraints and dynamics: constraint laws (as kinematical constraints) place restrictions on the types of states that the dynamics can affect and in this way explain why the latter are constrained in the way they are.

5.2.4 The Structural Account

On a first gloss, a structural explanation explains by virtue of specifying mathematical structure of the relevant sort. Dorato and Felline (2011: 161), for instance, claim that "... quantum theory provides a kind of *mathematical* explanation of the physical phenomena it is about. Following the available literature, we will refer to such explanations as *structural explanations*." They go on to give structural explanations of the Heisenberg uncertainty principle in terms of the mathematical structure of Fourier transformations, and the non-locality exhibited by entangled states in terms of the mathematical structure of the tensor product operation on a multi-particle Hilbert space. As examples of the available literature, they cite the following statements of Hughes (1989) and Clifton (1998):

> [A] structural explanation displays the elements of the models the theory uses and shows how they fit together. More picturesquely, it disassembles the black box, shows the working parts, and puts it together again. 'Brute facts' about the theory are explained by showing their connections with other facts, possibly less brutish. (Hughes, 1989: 198)

> We explain some feature B of the physical world by displaying a mathematical model of part of the world and demonstrating that there is a feature A of the model that corresponds to B, and is not explicit in the definition of the model. (Clifton, 1998: 7)

On first glance, the CPT and spin–statistics theorems might be thought of as providing structural explanations of CPT invariance and SSC insofar as they show how the mathematical structure associated with a set of principles, or a model of a set of axioms, demonstrates why states of a physical system must possess these properties. On second glance, however, it's not entirely clear how mathematical structure alone can explain. This concern is raised by Bueno and French (2012: 99) who argue that it is not enough for mathematical structure to carry explanatory weight that it just stands in a representational relation to physical structure; rather, the physical structure so-represented must exhibit dependence relations of the relevant sort. Thus, for instance, Lange (2013: 509) suggests that a "distinctively mathematical explanation" explains "by describing the framework inhabited by any possible causal relation." This framework is more fundamental than the causal relations that inhabit it insofar as it is supposed to show how the explanandum "... was inevitable to a stronger degree than could result from the causal powers bestowed by the possession of various properties" (Lange, 2013: 487). Moreover, this framework is supposed to work "by constraining what there could be" (Lange, 2013: 494). Let "the framework inhabited by any possible causal relation" be the space of kinematically possible states, and let causal relations be encoded in dynamics. One might then view the CPT and spin–statistics theorems as providing distinctively mathematical explanations of CPT invariance and SSC.

Alternatively, the relevant sort of dependence relations that Bueno and French request might be fleshed out counterfactually. Bokulich (2008: 226) suggests dropping Woodward's requirement that an explanation provide causal information and retaining only the requirement that it convey counterfactual information: "... while I shall adopt Woodward's account of explanation as the exhibiting of a pattern of counterfactual dependence, I will not construe this dependence narrowly in terms of the possible causal manipulations of the system." This motivates the following notion of a structural explanation:

> ... a structural explanation can be understood as one in which the explanandum is explained by showing how the (typically mathematical) structure of the theory itself limits what sort of objects, properties, states, or behaviors are admissible within the frame-work of that theory, and then showing that the explanandum is in fact a consequence of that structure. (Bokulich, 2011: 40)

Thus one might claim that the CPT and spin–statistics theorems provide structural explanations of CPT invariance and SSC by showing how the kinematically possible states of a physical system described by an RQFT are constrained (in terms of counterfactual dependencies) to those that possess these properties.[19] Moreover, Section 5.2.3 argued that the information that these theorems convey is not causal (under a charitable understanding of causal information). I think there is an essential problem with this structuralist understanding of CPT invariance and SSC, but before raising it, I would like to consider two examples: the "geometric" explanations of CPT invariance offered by Greaves (2010) and Swanson (2014).

Geometric Explanations of CPT Invariance as Structural Explanations

The structuralist appeal to constraints on kinematically possible states seems to underwrite the "geometric" explanations of CPT invariance given by Greaves (2010) and Swanson (2014). Both authors are concerned with providing an answer to the question, as posed by Swanson (2014: 62), "Why does this seemingly ad hoc assemblage of operations [i.e., C, P, and T] always yield a perfect symmetry?," but whereas Greaves addresses it in the Lagrangian formalism, Swanson employs the algebraic approach.

Greaves' (2010) strategy is to first adopt a particular interpretation of antimatter under which time reversal is identified with charge conjugation.[20] This allows Greaves to "change the subject" (2010: 38) from CPT to PT. An appeal is then made to Bell's (1955) classical PT theorem (see Section 4.2) to demonstrate that, if a classical field theory is characterized by fields that transform as restricted Lorentz invariant tensors, and by dynamical equations that are partial differential equations that express the vanishing of a local polynomial combination of the fields and their derivatives, then the fields are PT invariant. Greaves (2010: 43) interprets this as the claim that, subject to the theorem's antecedent conditions, "... there is no theory that makes essential use of a temporal orientation, over and above a Lorentzian metric and a total orientation." In slightly more detail, let (M, η_{ab}) be Minkowski spacetime, where M is a differential manifold and η_{ab} is the Minkowski metric. Greaves notes that the structures $(M, \eta_{ab}, \varepsilon_{abcd})$ and $(M, \eta_{ab}, \varepsilon_{abcd}, \tau)$, where ε_{abcd} is an orientation and τ is a temporal orientation, are invariant under the proper Lorentz group L_+ and the restricted Lorentz group $L_+^\uparrow$, respectively.[21] The classical PT theorem entails that a classical field theory that satisfies the antecedent conditions cannot be formulated in the latter structure, and rather must be formulated in the former. This is explained by the fact that "there is no tensor field that represents temporal orientation and no more, in the context of a flat Lorentzian metric and a total orientation" (Greaves, 2010: 43). Thus, according to Greaves, PT invariance, and hence (under the Bell/Feynman interpretation of antimatter) CPT invariance, is the consequence of a constraint imposed on a classical relativistic field theory. This constraint is geometric in origin, insofar as it encodes the incompatibility of certain types of geometric structures on the spacetime associated with the theory. To the extent that such a geometric constraint is a kinematical constraint, imposed independently of the specification of the dynamics of the theory, Greaves' explanation of CPT invariance by appeal to the classical PT theorem is a structural explanation.

Swanson's (2014) geometric explanation of CPT invariance appeals to the algebraic approach to RQFTs. As a C^*-algebra, a local von Neumann algebra of observables $\mathfrak{R}(\mathcal{O})$ admits a noncommutative product and an involution operation $*$: $\mathfrak{R}(\mathcal{O}) \rightarrow \mathfrak{R}(\mathcal{O})$. The latter defines a unique decomposition of the C^*-product into a symmetric component (the Jordan product), and an antisymmetric component which defines a Lie product. The Lie product subsequently defines an orientation on the state space associated with $\mathfrak{R}(\mathcal{O})$ (Alfsen and Shultz, 2001: 334). The involution operation also defines an isomorphism φ: $\mathfrak{R} \rightarrow \mathfrak{R}^{op}$, where $\mathfrak{R}^{op}$ is a copy of $\mathfrak{R}$ but with opposite C^*-multiplication. The map φ reverses the Lie product and thus reverses the orientation of the associated state space. Swanson views φ as a local reflection symmetry of $\mathfrak{R}(\mathcal{O})$, and interprets the algebraic CPT theorem as a procedure by means of which this local algebraic symmetry is extended, first to a local spacetime symmetry associated with a wedge algebra, and then to a global spacetime symmetry of the net of local algebras. A key part of this procedure is the recognition that the Lie product of a local algebra encodes symmetries in two ways: first, it encodes how observables generate continuous (spacetime) symmetries, and second, it encodes how unobservable field operators generate internal gauge symmetries (Swanson, 2014: 86).[22] Thus, according to Swanson, the Lie algebra structure of a local algebra explains why a transformation on its states that reverses their orientation involves both a spatiotemporal reflection and a charge conjugation. Under Swanson's view, then, CPT invariance is the consequence of a constraint imposed on an algebraic RQFT. This constraint is geometric in origin, insofar as it encodes the incompatibility of certain types of structures defined on the theory's state space; namely, that of a state space orientation, and temporal, spatial, and charge orientations.

As with Greaves' explanation, Swanson's can be described as structural, to the extent that it explains CPT invariance by virtue of an appeal to mathematical constraints imposed on a theory's state space that are independent of the specification of the theory's dynamics. Their explanations differ insofar as Greaves' account is formulated within the Lagrangian approach to field theories, whereas Swanson's is formulated in the algebraic approach. Moreover, as Swanson (2014: 98) notes, under Greaves' account CPT invariance is an essentially relativistic result, whereas for Swanson, it is an essentially relativistic and quantum mechanical result.

A Problem with Structuralist Accounts

It seems clear that the CPT and spin–statistics theorems come closest to providing structural explanations of CPT invariance and SSC. These theorems demonstrate how a set of principles constrains the kinematically possible states of a physical system described by an RQFT to possess these properties, irrespective of the dynamics such states obey. However, on closer inspection, a problem arises for such a structuralist interpretation.

Note first that the conceptually distinct approaches to the CPT and spin–statistics theorems suggest there are distinct, competing mathematical structures that can be associated with CPT invariance and SSC. On Bueno and French's (2012) view, it then seems strange to say there are distinct physical structures that underwrite CPT invariance and SSC. Under Lange's (2013) view, it seems strange to say there are distinct frameworks that any possible causal relation may inhabit. This complaint assumes that the different approaches to the CPT and spin–statistics theorems are not just conceptually distinct, but also distinct at the level of mathematical structure. I take this to be the case: naively, taken literally, the mathematical structures associated with Lagrangian field theory, axiomatic Wightman fields, von Neumann algebras, and *S*-matrices are all distinct. However, I grant that an argument could be made that, for a particular subclass of RQFTs, all of these approaches supervene on a common underlying mathematical structure. In particular, as noted in Section 1.4, there are models of non-interacting and unrealistic interacting RQFTs that do not face the *existence problem*. For such theories, one might attempt to identify a common underlying mathematical structure (perhaps encoded in the algebraic approach). However, the types of theories that seem to be of primary interest for explanatory accounts of CPT invariance and SSC are realistic interacting RQFTs, and interacting NQFTs and NQM. These are the types of theories that the empirical evidence indicates possess CPT invariance and SSC, and for these theories (at least currently) it will be hard to make the case for a common underlying mathematical structure.

The upshot of this discussion is that, for a subclass of physical systems that includes all non-interacting, and some unrealistic interacting systems, the CPT and spin–statistics theorems can be viewed as providing structural explanations of CPT invariance and SSC. For this subclass of systems, these theorems impose constraints on kinematically possible states in such a way as to limit these states to those that possess CPT invariance and SSC. These constraints are kinematical, insofar as they are robust under variations in the dynamics (as long as such variations remain within the subclass of systems). But one may ask, do these structural explanations underwrite an understanding of CPT invariance and SSC? The answer, arguably, is no. In the RQFT context, the systems of interest; those that make contact with empirical tests, lie outside the subclass of systems for which the

CPT and spin–statistics theorems provide structural explanations. In the NQM and NQFT contexts, the systems of interest likewise lie outside this subclass. The NQM and NQFT systems for which SSC in particular plays an essential explanatory role are interacting systems of BECs, superconductors, superfluids, and the like. To underwrite an understanding of CPT invariance and SSC then, we need to move beyond an appeal to the CPT and spin–statistics theorems.

5.3 What Would Explain CPT Invariance and the Spin–Statistics Connection?

CPT invariance and SSC can be derived in multiple ways for non-interacting (and some unrealistic interacting) RQFTs. These properties cannot be derived for realistic interacting RQFTs, and they cannot be derived for interacting NQFTs and interacting theories of NQM. This is problematic, since the experimental evidence for CPT invariance comes from realistic interacting RQFTs, whereas that for SSC comes from interacting NQFTs and NQM. I've just argued that these considerations entail that a full understanding of CPT invariance and SSC must go beyond an appeal to the CPT and spin–statistics theorems. But if not such an appeal, then *what* explains these properties? It will help to first distinguish two parts to this question:

(1) Why do systems described by interacting NQFTs and NQM exhibit SSC?
(2) Why do systems described by realistic interacting RQFTs exhibit SSC and CPT invariance?

Toward answering question (1), I'd like to consider an example discussed by Weatherall (2011). Weatherall is concerned with explaining the equivalence between inertial mass m_i and gravitational mass m_g in the context of Newtonian gravity.[23] In Weatherall's account, this equivalence is a general observational feature of the world that, while expressible in Newtonian gravity, is taken to be a brute empirical fact in that theory. On the other hand, while this equivalence is not expressible in GR, it can be shown to arise when one considers Newtonian gravity as a limiting case of GR. Thus, according to Weatherall,

> The explanatory demand is to show how, given some superseding theory, a general fact as expressed within one theory is really necessary or to be expected within the regime in which the old theory is successful... The explanatory work, then, is done by presenting the details of the relationship between the two theories. (Weatherall, 2011: 437)

In Weatherall's account, a major role is played by the fact that the explanandum ($m_i = m_g$) is not expressible in the superseding theory; in Weatherall's example, this is reflected in the fact that gravitational mass is not expressible in GR.

The weight of the explanation is thus carried primarily by the intertheoretic relation, as opposed to the derivation of the explanandum. In Weatherall's example, the equivalence between inertial and gravitational mass in Newtonian gravity is explained by showing how it can be derived in Newton–Cartan gravity, viewed as a limiting case of GR, and then demonstrating how Newtonian gravity can be recovered from Newton–Cartan gravity.[24] The framework for this explanation is given in Figure 5.1.

In GR, the *explanandum* (the equivalence between inertial and gravitational mass) is not expressible, whereas in Newton–Cartan gravity, it is derivable, and in Newtonian gravity, it is a brute fact. Relation A encodes a particular limit of GR that produces Newton–Cartan gravity, and relation B is underwritten by a theorem due to Trautman that describes the conditions under which Newtonian gravity can be recovered from Newton–Cartan gravity.[25] The relation between GR and Newtonian gravity in this explanation is important: GR is the superseding theory that is assumed to provide the more accurate description of the class of phenomena associated with the explanandum. On the other hand, "... there are still regimes in which Newtonian gravitation provides a satisfactory characterization of nature" (Weatherall, 2011: 429). Thus, according to Weatherall, the explanation provides an answer to the question "Given that we now believe GR to have superseded Newtonian theory, then why, insofar as Newtonian theory is a limiting case of GR, are inertial and gravitational mass equal in Newtonian theory?"

An analogous question can be posed with respect to the SSC in NQFTs and NQM (for the moment, I will put off a discussion of CPT invariance): "Given that we now believe RQFT to have superseded NQFT and NQM, then why, insofar as NQFT and NQM are limiting cases of RQFT, does SSC hold in NQFT and NQM?" Note that, as in Weatherall's example, the empirical evidence for the explanandum appears in the theories that have been superseded (NQFT and NQM). Note, furthermore, that the explanandum (SSC) can be derived in a limited version of the superseding theory, insofar as the spin–statistics theorem holds only for non-interacting (and some unrealistic interacting) RQFTs. Note, finally, that the full version of the superseding theory is realistic interacting RQFTs, and, as I shall attempt to argue below, SSC cannot be expressed in the latter. Thus, in analogy with Weatherall's example, the framework for an explanation of SSC in interacting NQFTs and NQM is given in Figure 5.2.

To complete this analogy, I have to accomplish two tasks. First, I have to make good on the assertion that SSC is not expressible in realistic interacting RQFTs,

General relativity ($m_i = m_g$ is not expressible)	$\xrightarrow{A}$	Newton–Cartan gravity ($m_i = m_g$ is derivable)	$\xrightarrow{B}$	Newtonian gravity ($m_i = m_g$ is a brute fact)

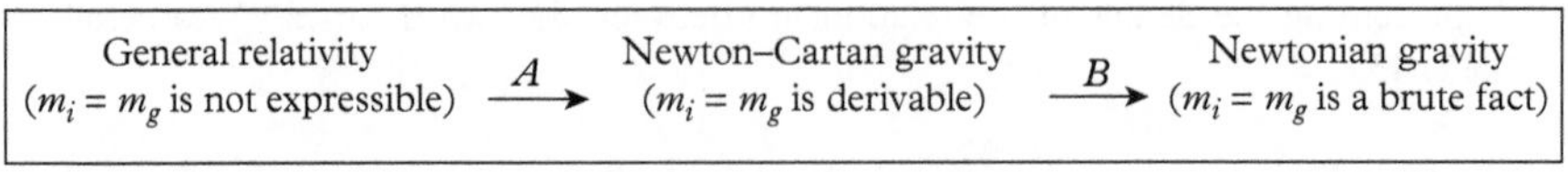

Figure 5.1 *Framework for an explanation of the equivalence between inertial and gravitational mass in Newtonian gravity.*

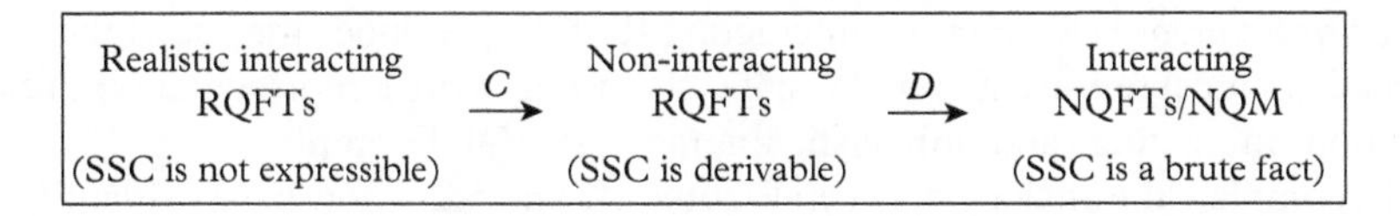

Figure 5.2 *Framework for an explanation of SSC in interacting NQFTs and NQM.*

and second, I have to articulate the nature of the relations C and D. With respect to the former task, in non-interacting RQFTs, SSC is expressed as a relation between spin states and states that obey statistics (call the latter "statistics states"): SSC says that half-integer spin states are states that obey Fermi–Dirac statistics, and integer spin states are states that obey Bose–Einstein statistics. Recall from Section 1.1.2 that in non-interacting RQFTs, spin states are represented by carriers of irreducible representations of the restricted Lorentz group $L_+^\uparrow$ or its double-covering group $SL(2, \mathbb{C})$. Statistics states are created or annihilated by field operators (or Fock space creation and annihilation operators) that satisfy the statistics–locality connection (StLC) of Section 1.1.2.[26] One way of demonstrating that this way of expressing SSC is unavailable in a realistic interacting RQFT is to recall the implications of Haag's theorem from Section 3.3.3. That theorem entails that a quantum field that satisfies the conjunction of the following conditions cannot be an interacting field:

(a) the field belongs to an irreducible representation of the equal-time canonical commutation relations;
(b) the field has a unique Euclidean-invariant vacuum state;
(c) the field is unitarily equivalent to a free field;
(d) the field is Poincaré-invariant.

In order for SSC to be attributed to the field, conditions (a), (b), and (d) must be upheld. Condition (a) allows statistics to be attributed to the field in the form of StLC. Conditions (b) and (d) allow spin to be attributed to the field by guaranteeing that it can be represented as a carrier of an irreducible representation of $L_+^\uparrow$ or $SL(2, \mathbb{C})$. This leaves condition (c). To deny it is to deny the possibility of constructing an S-matrix for the field (modulo the *existence problem*). Thus SSC cannot be attributed to an interacting field for which an S-matrix can be constructed. A few qualifications are in order at this point. First, Haag's theorem does not apply to non-interacting RQFTs or those unrealistic interacting RQFTs that are models of purist axioms; hence this argument against SSC expressibility in these latter theories fails. Second, what has been shown is that SSC, as expressed in a non-interacting RQFT, cannot be expressed in a realistic interacting RQFT associated with an S-matrix. This doesn't preclude the expression of SSC in realistic interacting RQFTs in a different way, but it's not that obvious what other ways there are of representing spin and statistics.[27] The general point is that the *existence problem* for realistic interacting RQFTs precludes the use of the types of mathematical representations of states that underwrite a representation of SSC that non-interacting (and unrealistic interacting) RQFTs employ.

To complete the analogy between Figures 5.1 and 5.2, I now have to identify the appropriate intertheoretic relations *C* and *D* in Figure 5.2. Relation *C* seems easy enough: given a realistic interacting RQFT, take the limit in which the interaction goes to zero. Relation *D* is a bit more complicated, but the groundwork for it has already been laid in Section 3.3.2. Note first that, as I've argued in Section 5.2, the different ways of deriving SSC in non-interacting RQFTs are conceptually and mathematically distinct, thus a single relation *D* will not be available. Rather, there will be one intertheoretic relation *D* for each approach to the spin–statistics theorem. As it turns out, this will not make any trouble.

Recall from Section 3.3.2 that a relation between non-interacting RQFTs and interacting Galilei-invariant QFTs (GQFTs) can be induced by a speed–space contraction of the Poincaré group. A speed–space contraction transforms the Poincaré group into the Galilei group and induces a transformation of irreducible representations of the former into irreducible representations of the latter. It also induces a transformation that takes spacelike intervals into spatial intervals at equal times. This subsequently induces transformations that take the relativistic versions of the locality constraints of *local commutativity* (LC), CD, *causality*, and *algebraic causality* into their non-relativistic versions. Section 3.3.2 argued that these facts underwrite the claim that a speed–space contraction defines a kinematical intertheoretic relation between RQFTs and GQFTs in each of the approaches to RQFTs and GQFTs reviewed in Sections 1.2 and 3.2. Moreover, Section 3.3.3 argued that a speed–space contraction also serves to underwrite an intertheoretic relation between *non-interacting* (and/or unrealistic interacting) RQFTs on the one hand, and *interacting* GQFTs on the other, and this makes it possible to "push down" essential properties of a non-interacting (or unrealistic interacting) RQFT to a corresponding interacting GQFT.

Thus in *non-interacting* RQFTs, SSC is an essential relation between spin states and statistics states. In (some) *interacting* GQFTs, SSC is a brute fact between spin states and statistics states. We'd like to explain this brute fact by an appeal to the derivation that grounds the essential relation in RQFTs, *and* in addition, an intertheoretic relation between non-interacting RQFTs and interacting GQFTs. The scaffolding for this explanation is represented in Figure 5.3. On the left we have an intertheoretic relation between non-interacting RQFTs and interacting GQFTs, and we'd like this relation to induce two relations in the figure on the right: one between relativistic and non-relativistic spin states, and the other between relativistic and non-relativistic statistics states. These induced relations will then allow us to "push down" the essential property of SSC from the *non-interacting* relativistic realm to the *interacting* non-relativistic realm.

A speed–space contraction of the Poincaré group induces a relation between integer and half-integer representations of the Poincaré group on the one hand, and integer and half-integer representations of the Galilei group on the other. Thus

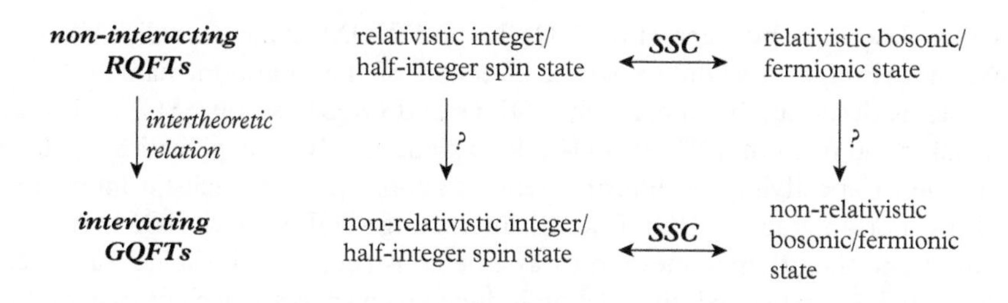

Figure 5.3 *The SSC and relations between RQFTs and GQFTs.*

it transforms relativistic spin states into non-relativistic spin states. Moreover, by transforming spacelike intervals into spatial intervals at equal times, it induces a relation between relativistic and non-relativistic versions of the StLC. Thus it transforms relativistic statistics states in all four approaches to the spin–statistics theorem reviewed in Chapter 1 into non-relativistic statistics states. It thus provides the glue to stitch together the right hand side of the diagram in Figure 5.3. Moreover, importantly, it stitches together the left hand side of Figure 5.3, too. There are true and double-valued representations of the Galilei group that describe *interacting* GQFTs. In such cases, a speed–space contraction represents an intertheoretic relation between *non-interacting* RQFTs and *interacting* GQFTs. In such cases, we have the diagram in Figure 5.4, which provides the framework within which to construct an explanation of SSC in interacting GQFTs (where P is the Poincare group, G is the Galilei group, and StLC is the statistics–locality connection).

Figure 5.4 shows that the general kinematical constraints shared by non-interacting RQFTs that are responsible for deriving SSC in each of the approaches to the spin–statistics theorem reviewed in Chapter 1 have non-relativistic counterparts under a speed–space contraction. These non-relativistic constraints explain SSC in interacting GQFTs by virtue of being the shadows of the relativistic constraints responsible for SSC in non-interacting RQFTs. Assumedly, these shadows hold equally for the finite-dimensional versions of GQFTs that take the forms of Galilei-invariant quantum mechanics (GQM).

At this point it might help to pause and take stock of the discussion so far. The first question posed at the beginning of this section was "Why do

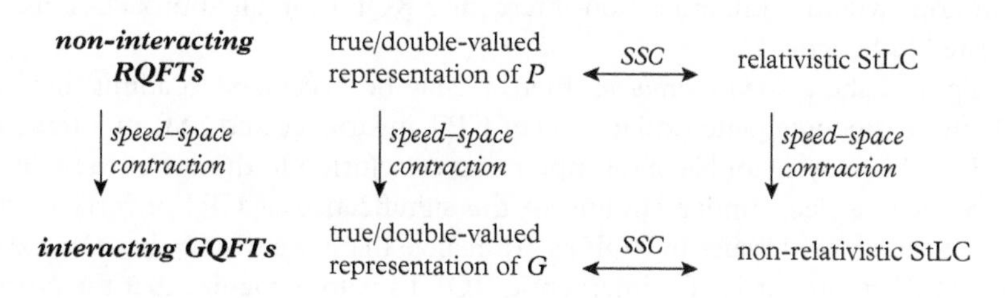

Figure 5.4 *Framework for an explanation of the SSC in interacting GQFTs.*

systems described by interacting NQFTs and NQM exhibit SSC?" We now have an answer to a slightly restricted version of this question, namely "Why do systems described by interacting GQFTs and GQM exhibit SSC?" This answer takes the form of a Weatherall-style explanation. This explanation explains by virtue of specifying the intertheoretic relations between realistic interacting RQFTs, non-interacting RQFTs, and interacting GQFTs and GQM. A similar answer to the original question cannot be constructed in the same way since, as Section 3.3.2 explained, the relation induced by a speed–space contraction between non-interacting RQFTs and interacting GQFTs cannot be extended to one between non-interacting RQFTs and interacting NQFTs in general, i.e., there are no similar relations between the Poincaré group and symmetry groups of classical spacetimes other than the Galilei group. This is not all that troubling insofar as the evidence for non-relativistic phenomena that exhibit SSC comes from interacting GQFTs and GQM. On the other hand, recall that in order to fully explain SSC, as well as CPT invariance, we also need an answer to question (2) posed at the beginning of this section, "Why do systems described by realistic interacting RQFTs exhibit SSC and CPT invariance?" An answer to this latter question is more complicated. Note first that it cannot be provided by a Weatherall-style explanation, insofar as there is no intertheoretic relation between non-interacting RQFTs and realistic interacting RQFTs that could carry the needed explanatory weight. More precisely, there is no relation that transforms a non-interacting RQFT into a realistic interacting RQFT. This is the *existence problem* of Section 1.4.3 faced by both pragmatist and purist approaches to RQFTs. The *existence problem* thus entails that the constraints that the CPT and spin–statistics theorems impose on the kinematically possible states of non-interacting RQFTs cannot be "pushed up" to constraints imposed on the kinematically possible states of realistic interacting RQFTs, in the same way that they can be "pushed down" to constraints imposed on the kinematically possible states of realistic interacting GQFTs and GQM.

Note that if there were such a relation, i.e., if there were a solution to the *existence problem*, then a Weatherall-style account of CPT invariance and SSC in interacting RQFTs would essentially reduce to a structural explanation. Such a relation would underwrite the claim that there is common structure shared between non-interacting and realistic interacting RQFTs, and that this structure explains CPT invariance and SSC in the latter. Assumedly, this common structure would not vary between the different approaches to formulating RQFTs, provided that a common structure underlies non-interacting RQFTs in all approaches, as was suggested in Section 5.2.4.

In light of the *existence problem*, then, it may be necessary to admit that currently there is no adequate explanation of CPT invariance and SSC in interacting RQFTs. This may seem like a disappointing conclusion to draw, but to maintain it is to adopt a deep understanding of the significance of CPT invariance and SSC to foundational issues in RQFTs. To realize that we currently have no explanation of these properties for interacting RQFTs is to recognize that the current status of interacting RQFTs is still very much up in the air, and that they are, in a non-trivial sense, very different types of theories than non-interacting RQFTs and interacting NQFTs and NQM.

5.4 Summary

What explains CPT invariance and the SSC? In this chapter, I've broken this question into three components. The first component asked "Do the CPT and spin–statistics theorems explain CTP invariance and SSC in RQFTs?" My answer was no: these theorems do not provide explanations of CPT invariance and SSC in RQFTs, at least under any of the standard accounts of scientific explanation (DN, unifying, causal, or structural). My argument was based on the facts that proofs of these theorems can be formulated in mathematically and conceptually distinct ways, that the theorems do not hold for realistic interacting RQFTs, and that the theorems do not hold for non-relativistic theories (NQFTs and NQM). This conclusion generated two additional questions, which constitute the second and third components to the original question. Since the evidence for CPT invariance comes from realistic interacting RQFTs, and much of the evidence for SSC comes from interacting GQFTs and GQM, an explanation of CPT invariance and SSC requires answers to the following questions:

(1) Why do systems described by interacting GQFTs and GQM exhibit SSC?
(2) Why do systems described by realistic interacting RQFTs exhibit SSC and CPT invariance?

My answer to question (1) was framed in terms of Weatherall's (2011) account of explanation. According to this account, realistic interacting RQFTs presumably offer the fundamental description of the phenomena that exhibit SSC, yet SSC is not expressible, nor derivable, in these theories. On the other hand, SSC is expressible and derivable in non-interacting (and some unrealistic interacting) RQFTs. Moreover, SSC is expressible, but not derivable in interacting GQFTs and GQM. An explanation of SSC in interacting GQFTs and GQM is provided by demonstrating how it arises in a limiting process as one goes from realistic interacting RQFTs to non-interacting RQFTs (by turning off interactions), and then to interacting GQFTs and GQM, by means of a speed–space contraction of the Poincaré group. In the latter limit, in each of the mathematically and conceptually distinct approaches to the spin–statistics theorem in non-relativistic RQFTs, the relativistic representations of spin and statistics are transformed into non-relativistic counterparts, as are the various relativistic locality constraints (LC, CD, *causality, algebraic causality*). As discussed in Chapter 3, a speed–space contraction transforms kinematical features of RQFTs into both kinematical and dynamical features of GQFTs. In this way, SSC as an essential property in non-relativistic RQFTs is pushed down to interacting GQFTs and GQM and thereby explained.

My answer to question (2) was less constructive: I suggested that currently there is no adequate explanation of CPT invariance and SSC in interacting RQFTs. To recognize this is to recognize that the *existence problem* for both purist and pragmatist approaches to interacting RQFTs has yet to be adequately solved.

NOTES

1. In the Schrödinger equation $i\partial_t\psi = E\psi$, one replaces the non-relativistic dispersion relation $E = \mathbf{p}^2/2m$ with the relativistic relation $E = (\mathbf{p}^2 + m^2)^{1/2}$. One can then show that solutions to the resulting equation do not have compact spatial support, i.e., they spread spatially with time (Strocchi, 2013: 3).
2. The problem of reconciling these principles can be made precise in the form of a theorem (Strocchi, 2013: 8, Proposition 2.2). Strocchi (2013: 10) views this theorem as a non-field-theoretic "counterpart" of the CPT theorem.
3. The basic trick amounts to the fact that in (2+1) dimensions, a boson can be turned into a fermion by coupling it to a Chern–Simons field.
4. Note that the mechanism of formation for a fractional quantum Hall liquid is different than in standard BEC formation. In the latter, the phase transition to the condensate is characterized by a spontaneously broken symmetry. In a fractional quantum Hall liquid, the phase transition is characterized, not by a symmetry, but by a topological order.
5. In addition, the derivation must be in the form of a sound deductive argument formulated in a first-order formal language, and the premises must possess empirical content. In the following, for charity's sake, I will set aside concerns having to do with the extent to which theories in mathematical physics are capable of first-order formulation. The original description of DN is given in Hempel and Oppenheim (1948).
6. In the RQFT case, admittedly, the interlocking nature of any given set of principles remains to be explicitly shown. This would require, in the Wightman axiomatic approach, for instance, that RLI, say, can be derived from the conjunction of the SC, *weak local commutativity*, and *CPT invariance*. More generally, one would like to show that the "derived" properties of CPT invariance and SSC can be used as assumptions in the derivation of any one of the alleged more fundamental principles, in conjunction with the others.
7. Scope is measured in terms of the number of conclusions that can be drawn from Σ. Simplicity is measured in terms of the size of Σ. Stringency is made precise in the following way. A systematization of K is a set of arguments Σ that instantiates a collection of *argument patterns*. An argument pattern consists of a schematic argument, a set of filling instructions, and a classification (the filling instructions explain how the schematic argument is to be interpreted, the classification identifies premises, conclusion, and inference

rules of the schematic argument). An argument pattern is more stringent than another if the first is harder to instantiate than the second.

8. If $E(K)$ were not unique, then the conjunction of Kepler's laws and Boyle's law might count as a legitimate explanation of Kepler's laws, although perhaps less unifying than, say, Galileo's law. Uniqueness of $E(K)$ entails, on the other hand, that the conjunction of Kepler's laws and Boyle's law is unexplanatory of Kepler's laws, which seems more appropriate.
9. Under this latter view, according to Strevens, "to explain a law is to have a kind of generalized understanding of the causes of instances of the law, what causalists often call an understanding of a mechanism." This view of mechanism underwrites Strevens' kairetic account of regularity explanation. Under another meaning of the term, a mechanism consists of a collection of entities and activities that are organized in such a way that they realize the regularity in question (see, e.g., Weber et al., 2013: 59 and references therein).
10. Note that the way I initially formulated the explanation of CPT invariance and SSC in terms of the CPT and spin–statistics theorems in claim (*) at the beginning of Section 5.2 might be interpreted as an instance of a metaphysical causal explanation. It takes the form of what Skow (2014: 446) refers to as an "in-virtue-of" explanation. It purports to explain why physical systems described by RQFTs possess CPT invariance and SSC by citing other more fundamental properties (RLI, the SC, etc.) that "ground" the properties in question. Thus it claims that CPT invariance and SSC obtain in a physical system in virtue of that system possessing some set of more fundamental properties. However, whereas Skow (2014: 447) dismisses in-virtue-of explanations of particular facts (events) as "obviously non-causal," Strevens (2008: 220) claims that they fail as causal explanations of regularities: "... the facts about causal influence are more or less fundamental physical facts, or so I suppose, and thus are suitable stopping points for understanding... By contrast, there is no relation of dependence between laws that can be read off the physics in the same way."
11. Some advocates of the causal history approach of Lewis appear to associate dynamical laws with causal histories in this way (Skow, 2014: 461).
12. This would entail that these theorems provide a causal explanation of CPT invariance and SSC, under Lewis' causal account of explanation, if it weren't for the fact that the latter account is restricted to explanations of particular events.
13. Saatsi and Pexton (2013) suggest that in some cases of Woodward-style explanations of regularities, explanatory power resides in the counterfactual content alone. Insofar as this type of Woodward-style explanation does not provide causal information about its explanandum, it is not a causal explanation. The appeal to counterfactual dependence alone as explanatory will be considered in the next section under structural accounts of explanation.
14. Saatsi's (2015) discussion of a type of non-causal geometric explanation of particular events similarly puts an emphasis on kinematical aspects of the event that are robust under variation of its dynamics.

15. Mullin and Blaylock (2003: 1224) examine a number of examples of purported fermion repulsion and bosonic attraction. The former include the virial correction to the ideal gas law, the electron degeneracy pressure of a white dwarf star, and the interaction between electrons in a diatomic hydrogen atom. An example of bosonic attraction is the behavior of trapped bosons when they condense to form a BEC.
16. Another attempt to make the concept of fermionic and bosonic effective forces respectable appears in the literature on Bohm's interpretation of quantum mechanics. Holland (1993: 284) suggests that Bohmian dynamics explains PEP insofar as, while the dynamical Bohmian trajectories of fermions never cross, those of bosons can. The causal story is then supplied by appeal to the pilot wave as the force that explains these dynamical results. However, Brown et al. (1999: 223) point out that Bohmian dynamics only secures the fact that, if two trajectories do not coincide at a given initial time, they never will, and that if two trajectories do coincide at a given initial time, then they will coincide at all future times.
17. It should perhaps be mentioned that one motivation for Skow's analysis is his critique of an example of a purportedly causal explanation that makes use of the PEP. Lewis (1986: 222) claims that an explanation of why a white dwarf stops collapsing by appealing to PEP is a causal explanation, even though no explicit cause can be identified; in particular, according to Lewis an appeal to a constraint on a state space is not an appeal to an explicit cause. Lewis, however, still thinks such an explanation is causal insofar as it provides "negative" information about the causal history of the star, i.e., in claiming no cause was present, we learn something about how the star evolved over time. Skow points out that an appeal to PEP can be given a causal interpretation: PEP prohibits electrons in the stellar matter from occupying the lowest energy state; hence as the star collapses, its electrons must fill up a range of energy states just above the ground state. This entails they possess a range of energies, and thus exert a pressure, according to the thermodynamic relation $P = -\partial U/\partial V$, where P is pressure, U is internal energy, and V is volume. It is this internal pressure that balances the gravitational pressure of collapse, thereby causing the star to stop collapsing (for a similar account, see Mullin and Blaylock, 2003: 1226). Note that the explanandum here (why a particular star stopped collapsing) is distinct from the explanandum that is of concern in this section, i.e., this section wants to know why the electrons in the stellar matter possess SSC, and hence obey PEP, in the first place, and not why, given that they do, how this explains why the star will stop collapsing.
18. "Whether to think we have here an instance of the second or the third possibility might depend on the relationship between the Pauli exclusion principle and the requirement of permutation invariance" (Skow, 2014: 462, footnote 21).
19. This also seems to be Railton's (1980: 350) notion of structural explanations as "explanations based upon laws that limit the possible states or

state-evolutions of systems." According to Railton, examples of constraint laws of this kind include the Heisenberg uncertainty principle, the PEP, Gibbs' phase rule, the first, second, and third laws of thermodynamics, conservation laws, and extremal principles.

20. Greaves (2010: 39) attributes this view to Bell (1955), and to Feynman's claim that "time reversal turns particles into antiparticles." The most literal understanding of this is to assume particles and antiparticles are identical except for the orientation of their momenta, but this violates the SC, as Swanson (2014: 90) observes (see also Baker and Halvorson, 2010: 118).
21. An orientation ε_{abcd} on Minkowski spacetime is a non-vanishing totally antisymmetric tensor field that serves to assign a handedness to tetrads (i.e., sets of four orthonormal vectors). A temporal orientation is a continuous choice of future light cone at each point. A spatial orientation ε_{abc} is a non-vanishing totally antisymmetric tensor field that assigns a handedness to orthonormal triads of spacelike vectors at each point. Elements of L_{+} leave ε_{abcd} invariant but reverse ε_{abc} and τ, whereas elements of $L_{+}^{\uparrow}$ leave ε_{abcd}, ε_{abc}, and τ invariant. Thus Greaves' interpretation of the classical PT theorem can be extended to the claim that, subject to the antecedent conditions, there is no theory that makes essential use of a temporal orientation *and* a spatial orientation, over and above a Lorentzian metric and an orientation.
22. The decomposition of the C^*-product of a local von Neumann algebra into symmetric (Jordan product) and antisymmetric (Lie product) components reflects the fact that "[quantum mechanical] physical variables play a dual role, as observables and as generators of transformation groups," as Alfsen and Schultz (2001: viii) state. It is with respect to the latter role that the Lie product encodes generators of spacetime symmetry groups and internal gauge groups.
23. Recall that inertial mass m_i is the constant of proportionality that appears in Newton's second law, $F = m_i a$. It measures the tendency of an object to obey Newton's first law. Gravitational mass m_g is a measure of the strength of the coupling of an object to the gravitational field, and appears in the definition of the Newtonian gravitational force $F = -m_g \partial \Phi$, where Φ is the Newtonian gravitational potential field.
24. GR is characterized by models of the form (M, g_{ab}), where M is a differential manifold and g_{ab} is a Riemannian metric on M that satisfies the Einstein equations. Newtonian gravity is characterized by models of the form $(M, h^{ab}, t_a, \nabla_a, \rho, \Phi)$, where h^{ab} and t_a are spatial and temporal metrics on M, ∇_a is a derivative operator, and ρ and Φ are scalar fields that represent the mass density and the Newtonian gravitational potential, respectively. These objects are required to satisfy $h^{ab}t_b = \nabla_c h^{ab} = \nabla_a t_b = 0$, $R^a_{\ bcd} = 0$, and $h^{ab}\nabla_a\nabla_b\Phi = 4\pi G\rho$ (Poisson equation), where $R^a_{\ bcd}$ is the curvature tensor defined by ∇_a. Newton–Cartan gravity is a geometricized version of Newtonian gravity characterized by models of the form $(M, h^{ab}, t_a, \nabla_a, \rho)$ which

satisfy $h^{ab}t_b = \nabla_c h^{ab} = \nabla_a t_b = 0$, $R^c_{abc} = R_{ab} = 4\pi G\rho t_a t_b$ (generalized Poisson equation), and $R^{ac}_{bd} = R^{ca}_{db}$.

25. With respect to relation A, the condition $R^{ac}_{bd} = R^{ca}_{db}$ on the curvature tensor of a model of Newton–Cartan gravity imposes a symmetry on the connection that makes it possible to recover it as the $c \to \infty$ limit of a (general relativistic) Riemannian connection (Malament, 1986). With respect to relation B, Trautman's recovery theorem requires an additional constraint on the curvature tensor of a model of Newton–Cartan gravity; namely, $R^{ab}_{cd} = 0$. This additional constraint is not necessary to underwrite relation A.
26. Recall from Section 1.2.4 that the algebraic proof of the spin–statistics theorem requires statistics to be encoded in a net of field algebras in the form of the constraint of *normal commutation relations*. This amounts to enforcing StLC on the elements of a field algebra.
27. A third point could be raised to the effect that not all realistic interacting RQFTs can be associated with S-matrices, e.g., asymptotically free theories like QCD. See footnote 40 in Chapter 1 for a discussion.

Conclusion

The preceding chapters have sought answers to the following questions:

(a) Why are CPT invariance and the spin–statistics connection (SSC) derivable properties in relativistic quantum field theories (RQFTs)?
(b) Why are CPT invariance and the SSC not derivable properties in non-relativistic quantum field theories (NQFTs)?
(c) Why are CPT invariance and the SSC not derivable properties in non-relativistic quantum mechanics (NQM)?
(d) What explains CPT invariance and the SSC?

Question (a) was addressed in Chapter 1 by comparing four distinct ways to formulate the CPT and spin–statistics theorems in RQFTs. These formulations were associated with four distinct formulations of RQFTs in general: the Wightman axiomatic approach, Weinberg's approach, the Lagrangian approach, and the algebraic approach. These approaches are both mathematically and conceptually distinct: taken at face value, the Wightman and Lagrangian approaches view RQFTs to be fundamentally about fields, yet they differ over the principles such fields must satisfy in order to derive CPT invariance and the SSC. Weinberg's approach views RQFTs to be about particles governed by the S-matrix, and subsequently identifies constraints that the S-matrix must satisfy in order to derive CPT invariance and the SSC. The algebraic approach commits itself to a net of local observable algebras realized as operators on a Hilbert space, and adopts yet another set of basic principles to underwrite CPT invariance and the SSC. Moreover, while the Wightman and algebraic approaches deem CPT invariance to be a necessary, but not sufficient, condition for the SSC, the Weinberg and Lagrangian approaches deem the SSC to be a necessary, but not sufficient, condition for CPT invariance. We also saw that the reason why CPT invariance and the SSC are derivable properties in RQFTs is not due *solely* to relativity: in the guise of *restricted Lorentz invariance* (RLI), relativity is neither necessary nor sufficient to derive these properties. It is not necessary, insofar as the algebraic approach does not employ it, and it is not sufficient insofar as all approaches employ additional principles in their derivations. Finally, we saw how these four distinct formulations of the CPT and spin–statistics theorems reflect two basic approaches to formulating RQFTs: pragmatist approaches and purist approaches. Both types of approach demonstrate that the CPT and spin–statistics theorems hold for non-interacting, and at best, unrealistic interacting RQFTs. Any extension of these theorems to realistic interacting RQFTs (and in particular, to the four-dimensional RQFTs that constitute the empirically successful standard model) must confront what I called the *existence problem*. The task for purists is to construct a model of an appropriate set of axioms that describes a realistic interacting RQFT. The task for pragmatists is to demonstrate that their preferred notion of existence holds for the realistic interacting theories of interest. The basic foundational problem that confronts interacting RQFTs is that these tasks have yet to be fully achieved.

Question (b) was addressed in part by considering the role that relativity plays in the CPT and spin–statistics theorems. In Chapter 2, we considered the relation between RLI and a kinematical locality constraint that appears in different forms in each of the approaches reviewed in Chapter 1: *local commutativity, cluster decomposition, causality*, and *algebraic causality*. We also considered the extent to which RLI is not necessary in the derivations of CPT invariance and the SSC by investigating the nature of the assumption of *modular covariance* in the algebraic approach, and related assumptions. Finally, a critique was provided of an influential claim made by Greenberg (2002) which maintains that if CPT invariance is violated in an interacting RQFT, then that theory also violates Lorentz invariance. We saw that this claim can be considered in either a pragmatist or a purist form, but either way it assumes a solution to the *existence problem*.

Chapter 3 further addressed question (b) by considering how NQFTs can be formulated in each of the approaches to RQFTs reviewed in Chapter 1. We saw that the reason why CPT invariance and the SSC fail to be derivable properties in NQFTs depends on which approach to NQFTs one adopts. Moreover, we described an intertheoretic relation, induced by a speed–space contraction of the Poincaré group, that transforms *kinematical* features of non-interacting RQFTs into both *kinematical* and *dynamical* features of interacting Galilei-invariant QFTs (GQFTs).

Question (c) was addressed in Chapter 4 by means of addressing the more general question of why CPT invariance and the SSC are not derivable properties outside the quantum field-theoretic framework. We found that the SSC can be derived for certain limited cases of systems described by NQM: spin-0 bosons under the assumptions of continuity and single-valuedness of the wavefunction (Peshkin, 2003a, 2003b, 2006), and two-dimensional multi-particle systems for which the total angular momentum operator $\mathcal{J}$ of any single-particle state is unitarily related to the total angular momentum operator j of the center of mass rest frame, with an extension, under even more restrictions, to three-dimensional systems (Kuckert, 2004: Kuckert and Mund, 2005). These results employ versions of the configuration space approach to deriving the SSC. This approach comes in an extrinsic form and an intrinsic form. We saw that the extrinsic form identifies particle indistinguishability with permutation invariance, and views the latter as a "single-valuedness" constraint on the system's wavefunction, defined on the simply-connected universal covering space of the system's multiply-connected classical configuration space. This approach obscures the understanding of permutation invariance as a constraint on a system's algebra of observables, and it makes use of a suspect concept of "single-valuedness" of the wavefunction. The intrinsic form of the configuration space approach, on the other hand, identifies particle indistinguishability with permutation invariance, and allows the latter to be conceived as a constraint on the algebra of observables. It does this by defining the wavefunction of the system in an intrinsic way on the multiply-connected classical configuration space (i.e., as a cross-section of an appropriate vector bundle defined on the latter). This formulation makes explicit the fact that the symmetrization postulate is a property only of certain types of cross-sections. Moreover, intrinsic approaches employ standard techniques in the quantization of topologically non-trivial configuration spaces and hence do not amount to non-standard formulations of quantum mechanics, as some critiques of extrinsic versions claim. On the other hand, the spin–statistics theorem in its complete generality has yet to be derived for NQM, even in the context of intrinsic configuration space approaches. We also saw how Sudarshan's (1968) approach to deriving the SSC in NQFTs deviates from standard formulations of the latter. Finally, we saw that, while CPT invariance is not inherently quantum mechanical, insofar as it can be derived in the context of the Lagrangian formulation of classical relativistic field theories, to do so requires employing a map between matter and antimatter states that arguably is not physically relevant in the classical field-theoretic context.

Question (d) was addressed in Chapter 5. We saw there that the CPT and spin–statistics theorems do not explain CPT invariance and the SSC, at least according to any of the accounts of explanation in the philosophy of science. We also saw that, nevertheless, an explanation of the SSC in interacting GQFTs and Galilei-invariant quantum mechanics (GQM) can be constructed. This explanation was based on two intertheoretic relations that took us from realistic interacting RQFTs (in which SSC cannot be expressed), to non-interacting RQFTs (in which SSC can be expressed and is an essential property), to interacting GQFTs and GQM (in which the majority of evidence for SSC can be found). As Weatherall (2011) suggests, this type of intertheoretic explanation may be a common feature of the way physicists explain phenomena.

On the other hand, this type of explanation cannot be appealed to in order to explain why realistic interacting RQFTs exhibit SSC and CPT invariance. If our understanding of these latter properties requires an explanation of their presence in these theories, then the *existence problem* currently precludes such an understanding.

Appendix
The Separating Corollary and the Axiomatic Proof of the Spin–Statistics Theorem

In the Wightman axiomatic proof of the spin–statistics theorem outlined in Section 1.2.1 of Chapter 1, one shows that if the wrong spin–statistics connection is assumed, then the fields are identically zero. This requires a "separating corollary," which plays an important role in the discussion of axiomatic NQFTs in Chapter 3. The goal of this appendix is to provide an account of this corollary and the role it plays in the axiomatic proof.

In the Wightman axiomatic treatment of RQFTs, one can associate to every bounded region $\mathcal{O}$ of Minkowski spacetime, the polynomial algebra of operators, bounded and unbounded, smeared with test functions with support in $\mathcal{O}$. This algebra is supposed to provide candidates for local measurements confined to $\mathcal{O}$. In the following I will restrict attention to von Neumann algebras consisting of all such bounded operators. One can show that Wightman axiom W4iii (the *spectrum condition* (SC); see Table 3.1 in Section 3.2.1) entails that the vacuum state is *cyclic* for any local (von Neumann) algebra of operators $\mathfrak{R}(\mathcal{O})$ associated with a spatiotemporal region $\mathcal{O}$ of spacetime. This means that the set of states $\{\phi\,|0\rangle\,;\phi \in \mathfrak{R}(\mathcal{O})\}$ generated by acting on the vacuum with any member of $\mathfrak{R}(\mathcal{O})$ is dense in $\mathcal{H}$.[1] This is the Reeh–Schlieder theorem (see, e.g., Streater and Wightman, 1964: 138). One next notes the following general result in the theory of operator algebras (see, e.g., Bratelli and Robinson, 1987: 85):

General result

Any cyclic vector for a von Neumann algebra $\mathfrak{R}$ is separating for its commutant $\mathfrak{R}'$.

The commutant $\mathfrak{R}'$ consists of operators that commute with all operators in $\mathfrak{R}$. Wightman axiom W3 (*relativistic local commutativity* (LC); see Table 3.1 in Section 3.2.1) entails that $\mathfrak{R}(\mathcal{O})' = \mathfrak{R}(\mathcal{O}')$, where $\mathcal{O}'$ is the *causal complement* of $\mathcal{O}$, defined as the set of all points *causally separated* (in this context, spacelike separated) from points in $\mathcal{O}$. In words: the commutant of a local algebra associated with a spacetime region $\mathcal{O}$ of Minkowski spacetime is the local algebra associated with the causal complement $\mathcal{O}'$ of $\mathcal{O}$. Thus, provided the causal complement of any region is non-empty, the general result entails that the vacuum is separating for any local algebra in Minkowski spacetime. We thus have the following *separating corollary* (Theorem 4-3 in Streater and Wightman, 1964: 139):

Separating corollary

Let $\mathfrak{R}(\mathcal{O})$ be a local algebra of operators associated with an open region $\mathcal{O}$ of spacetime. Suppose (i) the vacuum is cyclic for $\mathfrak{R}(\mathcal{O})$; (ii) the causal complement $\mathcal{O}'$ is non-empty; and (iii) relativistic local commutativity (W3) holds. Then the vacuum is *separating* for $\mathfrak{R}(\mathcal{O})$.

Separability of the vacuum means that, given any bounded region $\mathcal{O}$ of Minkowski spacetime, and any operator ϕ associated with $\mathcal{O}$, if ϕ annihilates the vacuum, $\phi\,|\,0\rangle = 0$, then it is identically zero, $\phi = 0$.

The axiomatic spin–statistics theorem (Streater and Wightman, 1964: 150) demonstrates that if the wrong spin–statistics connection is imposed on the fields $\phi(x)$, $\phi^{\dagger}(x)$, then, provided LC holds, the fields vanish. To obtain this result, we first encode statistics in the fields by assuming the *statistics–locality connection* (StLC) from Section 1.1.2, i.e., we assume bosonic fields commute and fermionic fields anti-commute (Streater and Wightman, 1964: 147). We next assume the *wrong* spin–statistics connection, i.e., we assume half-integer spin fields are bosonic and integer spin fields are fermionic. This entails half-integer-spin fields commute and integer-spin fields anti-commute at spacelike separated distances. Thus,

$$\begin{aligned} &[\phi(x), \phi^{\dagger}(y)]_{+} = 0 \quad \text{for } (x-y)^2 < 0 \text{ and integer spin } \phi, \\ &[\phi(x), \phi^{\dagger}(y)]_{-} = 0 \quad \text{for } (x-y)^2 < 0 \text{ and half integer spin } \phi. \end{aligned} \qquad \text{(NSpLC)}$$

This was called the *wrong spin–locality connection* (NSpLC) in Section 2.1.4. We now want to show that NSpLC entails that the fields annihilate the vacuum. We can then appeal to the separating corollary to argue that the fields must therefore vanish.

Note first that the vacuum expectation values of the fields are boundary values of complex Wightman functions in the following sense,

$$\begin{aligned} \langle 0|\,\phi(x)\phi^{\dagger}(y)\,|0\rangle &= \lim_{\eta\to 0} W(\zeta) \\ \langle 0\,|\,\phi^{\dagger}(y)\phi(x)\,|\,0\rangle &= \lim_{\eta\to 0} \hat{W}(-\zeta) \end{aligned}$$

where $\zeta = \xi - i\eta$, with $\xi, \eta \in \mathbb{R}$, $\xi = x - y$, and W, $\hat{W}$ are in general distinct complex Wightman functions of a single argument. The relations (NSpLC) induce the following relations between the corresponding Wightman functions,

$$W(\zeta) \pm \hat{W}(-\zeta) = 0 \qquad \text{(A.1)}$$

The PT condition of Section 1.2.1 entails that $\hat{W}$ satisfies

$$\hat{W}(\zeta) = (-1)^{\mathcal{J}}\,\hat{W}(-\zeta) \tag{A.2}$$

for $\mathcal{J}$ = (number of conjugate spinor indices in $\phi\phi^{\dagger}$) = (number of indices in ϕ) = even/odd if ϕ is integer/half-odd integer spin. So equation (A.2) is equivalent to

$$\hat{W}(\zeta) = \pm\hat{W}(-\zeta) \tag{A.3}$$

Combining equations (A.1) and (A.3) yields

$$W(\zeta) + \hat{W}(\zeta) = 0 \tag{A.4}$$

In the boundary limit $\eta \to 0$, this corresponds to $\langle 0|\,\phi(x)\phi^{\dagger}(y)|\,0\rangle + \langle 0\,|\,\phi^{\dagger}(-y)\,\phi(-x)|0\rangle = 0$, and this entails $\phi(x)\,|0\rangle = 0$. The separating corollary then entails that $\phi(x)$ must vanish. Thus the axiomatic spin–statistics theorem takes the schematic form (where NSSC denotes the wrong spin–statistics connection):

(i) [StLC & NSSC] ⇒ NSpLC
(ii) [(PT of Wightman functions) & NSpLC] ⇒ (fields annihilate the vacuum)
(iii) [LC & (fields annihilate the vacuum)] ⇒ (fields vanish)

where (iii) is the separating corollary (under the assumptions of a cyclic vacuum state and non-empty causal complements). Now note that a sufficient condition for PT invariance of Wightman functions is that the fields satisfy restricted Lorentz invariance (RLI) and the SC. We can use this result and combine (ii) and (iii) to obtain:

(ii′) [RLI & SC & LC & NSpLC] ⇒ (fields vanish)

Since StLC entails LC, (i) and (ii′) can be combined to produce

[RLI & SC & StLC & NSSC] ⇒ (fields vanish)

which is equivalent to

[RLI & SC & StLC] ⇒ [~NSSC ∨ (fields vanish)]

To obtain the form of the axiomatic spin–statistics theorem in Section 1.2.1 of Chapter 1, we have to assume that ~NSSC is equivalent to SSC (i.e., the failure of the wrong spin–statistics connection is just the spin–statistics connection). This essentially is the assumption that "... one puts aside the possibility of laws of statistics other than Bose–Einstein or Fermi–Dirac," as Streater and Wightman (1964: 147) state. We thus obtain

(RLI & SC & StLC] ⇒ (SSC for non-vanishing fields)

Note that entailment (i) above is not explicitly mentioned by Streater and Wightman (1964: 150). While they explicitly adopt StLC (1964: 147), they identity the "wrong connection of spin with statistics" with the wrong spin–locality connection NSpLC (1964: 150). Greenberg (1998: 146) points out that NSpLC "does not relate directly to particle statistics," but then goes on to claim "... for that reason this theorem should not be called the spin–statistics theorem." This appears to miss the implicit entailment (i) in Streater and Wightman's argument.

NOTE

1. This is to be distinguished from the cyclicity axiom (W4ii) (Table 3.2, Section 3.2.1), which is the requirement that the vacuum state be cyclic for the "global" algebra $\mathfrak{R}$, i.e., the von Neumann algebra of all operators defined on $\mathcal{H}$, as opposed to a local algebra $\mathfrak{R}(\mathcal{O})$ of operators defined only on test functions with support in $\mathcal{O}$.

References

Alfsen, E. and F. Shultz (2001) *State Spaces of Operator Algebras*. New York: Springer.

Allen, R. and A. Mondragon (2003) Comment on "Spin and statistics in nonrelativistic quantum mechanics: the spin-zero case". *Physical Review A* 68, 046101.

Amelino-Camelia, G., L. Freidel, J. Kowalski-Glikman, and L. Smolin (2014) Principle of relative locality. *Physical Review D* 84, 084010.

Annett, J. (2004) *Superconductivity, Superfluids and Condensates*. Oxford: Oxford University Press.

Arageorgis, A., J. Earman, and L. Ruetsche (2003) Fulling non-uniqueness and the Unruh effect: a primer on some aspects of quantum field theory. *Philosophy of Science* 70, 164–202.

Araki, H. (1999) *Mathematical Theory of Quantum Fields*. Oxford: Oxford University Press.

Araki, H., K. Hepp, and D. Ruelle (1962) On the asymptotic behavior of Wightman functions in space-like directions. *Helvetica Physica Acta* 35, 164–174.

Arntzenius, F. (2011) The CPT theorem. In: Callender, C. (ed.) *The Oxford Handbook of Philosophy of Time*. Oxford: Oxford University Press, pp.633–646.

Arntzenius, F. and H. Greaves (2009) Time reversal in classical electromagnetism. *British Journal for the Philosophy of Science* 60, 557–584.

Bacry, H. and J-M. Lévy-Leblond (1968) Possible kinematics. *Journal of Mathematical Physics* 9, 1605–1614.

Bain, J. (1999) Weinberg on QFT: demonstrative induction and underdeterminism. *Synthese* 117, 1–30.

Bain, J. (2004) Theories of newtonian gravity and empirical indistinguishability. *Studies in History and Philosophy of Modern Physics* 35, 345–376.

Bain, J. (2010) Relativity and quantum field theory. In: Petkov, V. (ed.) *Space, Time, and Spacetime—Physical and Philosophical Implications of Minkowski's Unification of Space and Time*. Berlin: Springer, pp.129–146.

Bain, J. (2011) Quantum field theories in classical spacetimes and particles. *Studies in History and Philosophy of Modern Physics* 42, 98–106.

Bain, J. (2014) Three principles of quantum gravity in the condensed matter approach. *Studies in History and Philosophy of Modern Physics* 46, 154–163.

Baker, D. (2009) Against field interpretations of quantum field theory. *British Journal for Philosophy of Science* 60, 585–609.

Baker, D. and H. Halvorson (2010) Antimatter. *British Journal for Philosophy of Science* 61, 93–121.

Baker, D., H. Halvorson, and N. Swanson (2014) The conventionality of parastatistics. *British Journal for Philosophy of Science*, axu018.

Balachandran, A., A. Daughton, Z. -C. Gu, R. Sorkin, G. Marmo, and A. Srivastava (1993) Spin–statistics theorems without relativity or field theory. *International Journal of Modern Physics A* 8, 2993–3044.

Bell, J. (1955) Time reversal in field theory. *Proceedings of the Royal Society of London A* 231, 479–495.

Belot, G. (1998) Understanding electromagnetism. *British Journal for the Philosophy of Science* 49, 531–555.
Benavides, C. and A. Reyes-Lega (2010) Canonical group quantization, rotation generators, and quantum indistinguishability. In: Ocampo, H., E. Pariguán, and S. Paycha (eds.) *Geometric and Topological Methods for Quantum Field Theory*. Cambridge: Cambridge University Press, pp.344–367.
Berger, M. (2011) Lorentz violation in top-quark production and decay. In: Kostelecky V.A. (ed.) *Proceedings of the 5th Meeting on CPT and Lorentz Symmetry*. Singapore: World Scientific, pp.179–183.
Berry, M. and J. Robbins (1997) Indistinguishability for quantum particles: spin, statistics and the geometric phase. *Proceedings of the Royal Society of London A* 453, 1771–1790.
Berry, M. and J. Robbins (2000a) Quantum indistinguishability: alternative constructions of the transported basis. *Journal of Mathematical Physics A* 33, L207–L214.
Berry, M. and J. Robbins (2000b) Quantum indistinguishability: spin–statistics without relativity or field theory? In: Hilborn, R. and G. Tino (eds.) *CP545, Spin–Statistics Connection and Commutation Relations*. American Institute of Physics, pp.3–15.
Bisognano, J. and Wichmann, E. (1975) On the duality condition for a Hermitian scalar field. *Journal of Mathematical Physics* 16, 985–1007.
Bisognano, J. and Wichmann, E. (1976) On the duality condition for quantum fields. *Journal of Mathematical Physics* 17, 303–321.
Bjorken, J. and S. Drell (1965) *Relativistic Quantum Fields*. New York: McGraw-Hill.
Bokulich, A. (2008) Can classical structures explain quantum phenomena? *British Journal for Philosophy of Science* 59, 217–235.
Bokulich, A. (2011) How scientific models can explain. *Synthese* 180, 33–45.
Bouatta, N. and J. Butterfield (2014) Renormalization for philosophers. *arXiv: 1406.4532v1*.
Bouatta, N. and J. Butterfield (2015) On emergence in gauge theories at the 't Hooft limit. *European Journal for Philosophy of Science* 5, 55–87.
Bourdeau, M. and R. Sorkin (1992) When can identical particles collide? *Physical Review D* 45, 687–696.
Bratelli, O. and D. Robinson (1987) *Operator Algebras and Quantum Statistical Mechanics*. New York: Springer.
Brennich, R. (1975) Deformation and contraction of Poincaré group representations. *Reports on Mathematical Physics* 8, 130–151.
Brown, H., E. Sjoqvist, and G. Bacciagaluppi (1999) Remarks on identical particles in de Broglie–Bohm theory. *Physics Letters A* 251, 229–235.
Buchholz, D. and S. Summers (1993) An algebraic characterization of vacuum states in Minkowski space. *Communications in Mathematical Physics* 155, 449–458.
Buchholz, D. and S. Summers (2004) An algebraic characterization of vacuum states in Minkowski space. III. Reflection maps. *Communications in Mathematical Physics* 246, 625–641.
Buchholz, D., M. Florig, and S. Summers (1999) An algebraic characterization of vacuum states in Minkowski space. II. Continuity aspects. *Letters in Mathematical Physics* 49, 337–350.
Buchholz, D., O. Dreyer, M. Florig, and S. Summers (2000) Geometric modular action and spacetime symmetry groups. *Reviews in Mathematical Physics* 12, 475–560.
Bueno, O. and S. French (2012) Can mathematics explain physical phenomena? *British Journal for Philosophy of Science* 63, 85–113.

Burgess, C.P. (2007) An introduction to effective field theory. *Annual Review of Nuclear and Particle Science* 57, 329–367.

Burgoyne, N. (1958) On the connection of spin and statistics. *Nuovo Cimento* 8, 607–609.

Caulton, A. (2013) Discerning "indistinguishable" quantum systems. *Philosophy of Science* 80, 49–72.

Chaichian, M., A. Dolgov, V. Novikov, and A. Tureanu (2011) CPT violation does not lead to violation of Lorentz invariance and vice versa. *Physics Letters B* 699, 177–180.

Christian, J. (1997) Exactly soluble sector of quantum gravity. *Physical Review D* 56, 4844–4877.

Clifton, R. (1998) Scientific explanation in quantum theory. Available at: <http://philsci-archive.pitt.edu/archive/91/>.

Dadashev, L.A. (1985) Axiomatics of Galileo-invariant quantum field theory. *Institute of Mathematics, Azerbaidzhan SSR Academy of Sciences*, 903–914. Translated from *Teoreticheskaya i Matematicheskaya Fizika* 64, 383–399.

Derezinski, J. and C. Gerard (1997) *Scattering Theory of Classical and Quantum N-Particle Systems*. Berlin: Springer.

Doplicher, S. and J. Roberts (1990) Why there is a field algebra with a compact gauge group describing the superselection structure in particle physics. *Communications in Mathematical Physics* 131, 51–107.

Doplicher, S., R. Haag, and J. Roberts (1971) Local observables and particle statistics I. *Communications in Mathematical Physics* 23, 199–230.

Doplicher, S., R. Haag, and J. Roberts (1974) Local observables and particle statistics II. *Communications in Mathematical Physics* 35, 49–85.

Dorato, M. and L. Felline (2011) Scientific explanation and scientific structuralism. In: Bokulich, P. and A. Bokulich (eds.) *Scientific Structuralism*. Dordrecht: Springer, pp.161–176.

Duck, I. and E. Sudarshan (1997) *Pauli and the Spin–Statistics Theorem*. Singapore: World Scientific.

Duck, I. and E. Sudarshan (1998) Toward an understanding of the spin-statistics theorem. *American Journal of Physics* 66, 284–303.

Duncan, A. (2012) *The Conceptual Framework of Quantum Field Theory*. Oxford: Oxford University Press.

Dütsch, M. and J. Gracia-Bondía (2012) On the assertion that PCT violation implies Lorentz non-invariance. *Physics Letters B* 711, 428–433.

Earman, J. (2010) Understanding permutation invariance in quantum mechanics. *Unpublished preprint*.

Earman, J. (2011) The Unruh effect for philosophers. *Studies in History and Philosophy of Modern Physics* 42, 81–97.

Earman, J. and D. Fraser (2006) Haag's theorem and its implications for the foundations of quantum field theory. *Erkenntnis* 64, 305–344.

Earman, J. and G. Valente (2014) Relativistic causality in algebraic quantum field theory. *International Studies in the Philosophy of Science* 28, 1–48.

Feynman, R. (1965) *The Feynman Lectures on Physics, Vol. 3*. Reading, MA: Addison-Wesley.

Fierz, M. (1939) Über die relativistische Theorie kräftefeier Teilchen mit beliebigem Spin. *Helvetica Physica Acta* 12, 3–37.

Finkelstein, J. and D. Rubinstein (1968) Connection between spin, statistics, and kinks. *Journal of Mathematical Physics* 9, 1762–1779.

Forte, S. (2007) Spin in quantum field theory. *Lecture Notes in Physics* 712, 67–94.

Fraser, D. (2006) *Haag's theorem and the interpretation of quantum field theories with interactions*. Ph.D. Thesis, University of Pittsburgh, USA.

Fraser, D. (2008) The fate of "particles" in quantum field theories with interactions. *Studies in History and Philosophy of Modern Physics* 39, 841–859.

Fraser, D. (2009) Quantum field theory: underdetermination, inconsistency, and idealization. *Philosophy of Science* 76, 536–567.

Fraser, D. (2011) How to take particle physics seriously: a further defense of axiomatic quantum field theory. *Studies in History and Philosophy of Modern Physics* 42, 126–135.

Friedman, M. (1974) Explanation and scientific understanding. *Journal of Philosophy* 71, 5–19.

Gell-Mann, M. and F. Low (1951) Bound states in quantum field theory. *Physical Review* 84, 350–354.

Geroch, R. (2011) Faster than light? *arXiv:1005.1614v1 [gr-qc]*.

Glaser, V., H. Lehman, and W. Zimmerman (1957) Field operators and retarded functions. *Nuovo Cimento* 6, 1122–1128.

Gorelik, G. and V. Frenkel (1994) *Matvei Petrovich Bronstein and Soviet Theoretical Physics in the Thirties*. Basel: Birkhäuser.

Greaves, H. (2008) *Spacetime symmetries and the CPT theorem*. Ph.D. Thesis, Rutgers University, USA.

Greaves, H. (2010) Towards a geometrical understanding of the CPT theorem. *British Journal of Philosophy of Science* 61, 27–50.

Greaves, H. and T. Thomas (2014) On the CPT theorem. *Studies in History and Philosophy of Modern Physics* 45, 46–65.

Greenberg, O. (1998) Spin–statistics, spin–locality, and TCP: three distinct theorems. *Physics Letters B* 416, 144–149.

Greenberg, O. (2002) CPT violation implies violation of Lorentz invariance. *Physical Review Letters* 89, 231602.

Greenberg, O. (2006a) Covariance of time-ordered products implies local commutativity of fields. *Physical Review D* 73, 087701.

Greenberg, O. (2006b) Why is CPT fundamental? *Foundations of Physics* 36, 1535–1553.

Guido, D. and R. Longo (1995) An algebraic spin and statistics theorem. *Communications in Mathematical Physics* 172, 517–533.

Haag, R. (1996) *Local Quantum Physics*. 2nd. ed. Berlin: Springer.

Hagar, A. (2009) Minimal length in quantum gravity and the fate of Lorentz invariance. *Studies in History and Philosophy of Modern Physics* 40, 259–267.

Hagen, C. (2004) Spin and statistics in Galilean covariant field theory. *Physical Review A* 70, 012101.

Halvorson, H. (2001) Reeh–Schlieder defeats Newton–Wigner: on alternative localization schemes in relativistic quantum field theory. *Philosophy of Science* 68, 111–133.

Halvorson, H. and M. Müger (2006) Algebraic quantum field theory. In: Butterfield, J. and J. Earman (eds.) *Philosophy of Physics*. Amsterdam: Elsevier, pp.731–922.

Harrison, J. and J. Robbins (2000) A group-theoretic approach to constructions of non-relativistic spin–statistics. In: Hilborn, R. and G. Tino (eds.) *CP545, Spin–Statistics Connection and Commutation Relations*. American Institute of Physics, 67–71.

Harrison, J. and J. Robbins (2004) Quantum indistinguishability from general representations of $SU(2\,n)$. *Journal of Mathematical Physics* 45, 1332–1358.

Hempel, C. and P. Oppenheim (1948) Studies in the logic of explanation. *Philosophy of Science* 15, 135–175.

Holland, P. (1993) *The Quantum Theory of Motion*. Cambridge: Cambridge University Press.

Holland, P. and H. Brown (2003) The non-relativistic limits of the Maxwell and Dirac equations: the role of Galilean and gauge invariance. *Studies in History and Philosophy of Modern Physics* 34, 161–187.

Horuzhy, S.S. (1990) *Introduction to Algebraic Quantum Field Theory*. Dordrecht: Kluwer Academic Publishers.

Huggett, N. and T. Imbo (2009) Indistinguishability. In: Greenberger, D., K. Hentschel, and F. Weinert (eds.) *Compendium of Quantum Physics*. Berlin: Springer, pp.311–317.

Hughes, R. I. G. (1989) Bell's theorem, ideology, and structural explanation. In: Cushing, J. and E. McMullin (eds.) *Philosophical Consequences of Quantum Theory: Reflections on Bell's Theorem*. Notre Dame: University of Notre Dame Press, pp.195–207.

Inonu, E. and E. Wigner (1953) On the contraction of groups and their representations. *Proceedings of the National Academy of Sciences* 39, 510–524.

Isham, C. (1984) Topological and global aspects of quantum theory. In: DeWitt, B. and R. Stora (eds.) *Relativity, Groups and Topology II*. Amsterdam: Elsevier, pp.1059–1290.

Jost, R. (1957) Eine Bemerkung zum CTP Theorem. *Helvetica Physica Acta* 30, 409–416.

Kaku, M. (1993) *Quantum Field Theory*. Oxford: Oxford University Press.

Kitcher, P. (1989) Explanatory unification and the causal structure of the world. In: Kitcher, P. and W. Salmon (eds.) *Scientific Explanation*. Minneapolis: University of Minnesota Press, pp.410–505.

Kostelecky, V.A. (ed.) (2011) *Proceedings of the Fifth Meeting on CPT and Lorentz Symmetry*. Singapore: World Scientific.

Kuckert, B. (1995) A new approach to spin & statistics. *Letters in Mathematical Physics* 35, 319–331.

Kuckert, B. (2004) Spin and statistics in nonrelativistic quantum mechanics, I. *Physics Letters A* 322, 47–53.

Kuckert, B. (2005) Spin, statistics, and reflections I. Rotation invariance. *Annales Henri Poincare* 6, 849–862.

Kuckert, B. (2007) The classical and quantum roots of Pauli's spin–statistics relation. *Lecture Notes in Physics* 718, 207–228.

Kuckert, B. and R. Lorenzen (2007) Spin, statistics, and reflections II. Lorentz invariance. *Communications in Mathematical Physics* 269, 809–831.

Kuckert, B. and J. Mund (2005) Spin & statistics in nonrelativistic quantum mechanics, II. *Annalen der Physik (Leipzig)* 14, 309–311.

Laidlaw, M. and C. DeWitt (1974) Feynman functional integrals for systems of indistinguishable particles. *Physical Review D* 3, 1375–1378.

Landsman, N. P. (2013) Quantization and superselection sectors III: multiply connected spaces and indistinguishable particles. *arXiv:1302.3637v2 [math-ph]*.

Lange, M. (2013) What makes a scientific explanation distinctively mathematical? *British Journal for Philosophy of Science* 64, 485–511.

Lehmann, H., K. Symanzik, and W. Zimmermann (1955) Zur Formulierung quantisierter Feldtheorien. *Nuovo Cimento* 1, 205–225.

Leinaas, J. and J. Myrheim (1977) On the theory of identical particles. *Il Nuovo Cimento* 37, 1–23.

Lévy-Leblond, J.-M. (1965) Une nouvelle limite non-relativiste du groupe de Poincaré. *Annales de l'Institute Henri Poincaré A*, 1–12.

Lévy-Leblond, J.-M. (1967) Galilean quantum field theories and a ghostless Lee model. *Communications in Mathematical Physics* 4, 157–176.

Lévy-Leblond, J.-M. (1971) Galilei group and Galilean invariance. In: Loebl, E. (ed.), *Group Theory and its Applications, Vol. II.* New York: Academic Press, pp.221–299.

Lewis, D. (1986) Causal explanation. In: Lewis, D. (ed.) *Philosophical Papers, Vol. 2.* New York: Oxford University Press, pp.325–347.

Liberati, S. (2013) Tests of Lorentz invariance: a 2013 update. *Classical and Quantum Gravity* 30, 133001.

Lorenzen, R. (2007) *From spin groups and modular P_1CT symmetry to covariant representations and the spin–statistics theorem.* Ph.D. Dissertation, University of Hamburg, Germany.

Lüders, G. and B. Zumino (1958) Connection between spin and statistics. *Physical Review* 110, 1450–1453.

Maggiore, M. (2005) *A Modern Introduction to Quantum Field Theory.* Oxford: Oxford University Press.

Malament, D. (1986) Newtonian gravity, limits, and the geometry of space. In: Colodny, R. (ed.) *From Quarks to Quasars: Philosophical Problems of Modern Physics.* Pittsburgh, PA: Pittsburgh University Press, pp.181–201.

Manohar, A. (1997) Effective field theories. *Lecture Notes in Physics* 479, 311–362.

Massimi, M. and M. Redhead (2003) Weinberg's proof of the spin–statistics theorem. *Studies in History and Philosophy of Modern Physics* 34, 621–650.

McCabe, G. (2007) *The Structure and Interpretation of the Standard Model.* Amsterdam: Elsevier.

Mickelsson, J. (1984) Geometry of spin and statistics in classical and quantum mechanics. *Physical Review D* 30, 1843–1845.

Morandi, G. (1992) *The Role of Topology in Classical and Quantum Physics.* Berlin: Springer-Verlag.

Morgan, J. (2004) Spin and statistics in classical mechanics. *American Journal of Physics* 72, 1408–1417.

Mullin, W. and G. Blaylock (2003) Quantum statistics: is there an effective fermion repulsion or boson attraction? *American Journal of Physics* 71, 1223–1231.

Oksak, A. I. and I. T. Todorov (1968) Invalidity of TCP-theorem for infinite-component fields. *Communications in Mathematical Physics* 11, 125–130.

Olive, K. and Particle Data Group (2014) Review of particle physics. *Chinese Physics C* 38, 090001.

Papadopoulos, N. and A. Reyes-Lega (2010) On the geometry of the Berry–Robbins approach to spin-statistics. *Foundations of Physics* 40, 829–851.

Papadopoulos, N., M. Paschke, A. Reyes, and F. Scheck (2004) The spin–statistics relation in nonrelativistic quantum mechanics and projective modules. *Annales Mathematiques Blaise Pascal* 11, 205–220.

Pauli, W. (1940) The connection between spin and statistics. *Physical Review* 58, 716–722.

Peshkin, M. (2003a) Spin and statistics in nonrelativistic quantum mechanics: the spin-zero case. *Physical Review A* 67, 042102.

Peshkin, M. (2003b) Reply to "Comment on 'Spin and statistics in nonrelativistic quantum mechanics: the spin-zero case'". *Physical Review A* 68, 046102.

Peshkin, M. (2006) Spin-zero particles must be bosons: a new proof within nonrelativistic quantum mechanics. *Foundations of Physics* 36, 19–29.

Peskin, M. and D. Schroeder (1995) *An Introduction to Quantum Field Theory*. New York: Addison-Wesley.

Polchinski, J. (1984) Renormalization and effective Lagrangians. *Nuclear Physics B* 231, 269–295.

Psillos, S. (2007) Past and contemporary perspectives on explanation. In: Kuipers, T. (ed.) *Handbook of the Philosophy of Science: General Philosophy of Science – Focal Issues*. Amsterdam: Elsevier, pp.97–173.

Puccini, G. and H. Vucetich (2004a) Axiomatic foundations of Galilean quantum field theories. *Foundations of Physics* 34, 263–295.

Puccini, G. and H. Vucetich (2004b) Possibility of obtaining a non-relativistic proof of the spin–statistics theorem in the Galilean frame. *arXiv:quant-ph/0407208v3*.

Puccini, G. and H. Vucetich (2005) Note on possibility of obtaining a non-relativistic proof of the spin–statistics theorem in the Galilean frame. *arXiv:quant-ph/0502048v1*.

Railton, P. (1980) *Explaining explanation: a realist account of scientific explanation and understanding*. Ph.D. Dissertation, Princeton University, USA.

Rebenko, A. (2012) *Theory of Interacting Quantum Fields*. Berlin: De Gruyter.

Requardt, M. (1982) Spectrum condition, analyticity, Reeh–Schlieder and cluster properties in non-relativistic Galilei-invariant quantum theory. *Journal of Physics A* 15, 3715–3723.

Reyes-Lega, A. (2011) On the geometry of quantum indistinguishability. *Journal of Physics A* 44, 1–20.

Reyes-Lega, A. and C. Benavides (2010) Remarks on the configuration space approach to spin-statistics. *Foundations of Physics* 40, 1004–1029.

Rivasseau, V. (1991) *From Perturbative to Constructive Renormalization*, Princeton: Princeton University Press.

Rivasseau, V. (2003) An introduction to renormalization. *Poincaré Seminar 2002*, 139–177.

Roberts, B. (2014a) Three merry roads to T-violation. *Studies in History and Philosophy of Modern Physics* <http://dx.doi.org/10.1016/j.shpsb.2014.08.003>.

Roberts, B. (2014b) Three myths about time reversal in quantum theory, <http://personal.lse.ac.uk/ROBERT49/pdf/ThreeMyths.pdf>.

Saatsi, J. (2015) On explanations from "geometry of motion", <juhasaatsidotorg.files.wordpress.com/2015/01/geometry-of-motion.pdf>.

Saatsi, J. and M. Pexton (2013) Reassessing Woodward's account of explanation: regularities, counterfactuals, and noncausal explanations. *Philosophy of Science* 80, 613–624.

Schwartz, M. (2013) *Quantum Field Theory and the Standard Model*. Cambridge: Cambridge University Press.

Schwinger, J. (1951) The theory of quantized fields. I. *Physical Review* 82, 914–927.

Schwinger, J. (1958) Spin, statistics, and the TCP theorem. *Proceedings of the National Academy of Sciences* 44, 223–228.

Shaji, A. (2009) Sudarshan's non-relativistic approach to the spin–statistics theorem. *Journal of Physics: Conference Series* 196, 012013.

Shaji, A. and E. Sudarshan (2003) Non-relativistic proofs of the spin–statistics connection. *arxiv:quant-phy/0306033v2*.

Skow, B. (2014) Are there non-causal explanations (of particular events)? *British Journal for Philosophy of Science* 65, 445–467.

Sozzi, M. (2008) *Discrete Symmetries and CP Violation: From Experiment to Theory*. Oxford: Oxford University Press.

Spekkens, R. (2013) The paradigm of kinematics and dynamics must yield to causal structure. *arXiv:1209.0023v1*.

Stachel, J. (2003) A brief history of space–time. In: Ciufolini, I. and D. Dominici (eds.) *2001: A Relativistic Space Odyssey, Experiments and Theoretical Viewpoints on General Relativity and Quantum Gravity*. Singapore: World Scientific, pp.15–34.

Sterman, G. (1993) *An Introduction to Quantum Field Theory*. Cambridge: Cambridge University Press.

Streater, R. (1967) Local field with the wrong connection between spin and statistics. *Communications in Mathematical Physics* 5, 88–98.

Streater, R. and A. Wightman (1964) *PCT, Spin and Statistics, and All That*. Princeton, NJ: Princeton University Press.

Strevens, M. (2008) *Depth: An Account of Scientific Explanation*. Cambridge, MA: Harvard University Press.

Strocchi, F. (2013) *An Introduction to Non-Perturbative Foundations of Quantum Field Theory*. Oxford: Oxford University Press.

Sudarshan, E. (1968) The fundamental theorem on the relation between spin and statistics. *Proceedings of the Indian Academy of Sciences A* 67, 284–293.

Sudarshan, E. and I. Duck (2003) What price the spin–statistics theorem? *Pramana* 61, 645–653.

Sudarshan, E. and A. Shaji (2004) Note on non-relativistic proof of the spin–statistics connection in the Galilean frame. *arxiv:quant-ph/0409205v1*.

Summers, S. (2006) Tomita–Takesaki modular theory. In: Françoise, J., G. Naber, and T. Tsun (eds.) *Encyclopedia of Mathematical Physics*. Elsevier, pp.251–257.

Summers, S. (2011) Yet more ado about nothing: the remarkable relativistic vacuum state. In: Halvorson, H. (ed.) *Deep Beauty: Understanding the Quantum World Through Mathematical Innocation*. Cambridge: Cambridge University Press, pp.317–341.

Summers, S. (2012) A perspective on constructive quantum field theory. *arXiv: 1203.3991v1 [math-ph]*.

Summers, S. and R. White (2003) On deriving space–time from quantum observables and states. *Communications in Mathematical Physics* 237, 203–220.

Swanson, N. (2014) *Modular theory and spacetime structure in QFT*. Ph.D. Dissertation, Princeton University, USA.

Tscheuschner, R.D. (1989) Topological spin–statistics relation in quantum field theory. *International Journal of Theoretical Physics* 28, 1269–1310.

Tscheuschner, R.D. (1990) Erratum and comment: topological spin–statistics relation in quantum field theory. *International Journal of Theoretical Physics* 29, 1437–1438.

Tscheuschner, R.D (1991) Coinciding versus noncoinciding: is the topological spin–statistics theorem already proven in quantum mechanics? *Journal of Mathematical Physics* 32, 749–752.

Verch, R. (2006) Vacuum fluctuations, geometric modular action and relativistic quantum information theory. *Lecture Notes in Physics*, 702, 133–162.

Wallace, D. (2006) In defence of naivete: the conceptual status of Lagrangian quantum field theory. *Synthese* 151, 33–80.

Wallace, D. (2009) QFT, antimatter, and symmetry. *Studies in History and Philosophy of Modern Physics* 40, 209–222.

Wallace, D. (2011) Taking particle physics seriously: a critique of the algebraic approach to quantum field theory. *Studies in History and Philosophy of Modern Physics* 42, 116–125.

Weatherall, J. (2011) On (some) explanations in physics. *Philosophy of Science* 78, 421–447.

Weatherall, J. (2014) Inertial motion, explanation, and the foundations of classical spacetime theories. In: Lehmkuhl, D., G. Schiemann, and E. Scholz (eds.) *Towards a Theory of Spacetime Theories (Einstein Studies series)*, Boston: Birkhauser, *arXiv:1206.2980v1.*

Weber, E., J. van Bouwel, and L. de Vreese (2013) *Scientific Explanation.* Dordrecht: Springer.

Weinberg, S. (1964) Feynman rules for any spin. *Physical Review B* 133, 1318–1332.

Weinberg, S. (1979) Ultraviolet divergences in quantum theories of gravitation. In: Hawking, S. and W. Israel (eds.) *General Relativity: An Einstein Centenary Survey.* Cambridge: Cambridge University Press, pp.790–901.

Weinberg, S. (1995) *The Quantum Theory of Fields, Vol. 1.* Cambridge: Cambridge University Press.

Weinberg, S. (1996) *The Quantum Theory of Fields, Vol. 2.* Cambridge: Cambridge University Press.

Weinstein, S. (2006) Superluminal signaling and relativity. *Synthese* 148, 381–399.

Wightman, A. (1956) Quantum field theory in terms of vacuum expectation values. *Physical Review* 101, 860–866.

Wightman, A. (1999) Review of I. Duck and E.C.G. Sudarshan's *Pauli and the Spin–Statistics Theorem. American Journal of Physics* 67, 742–746.

Wightman, A. (2000) The spin–statistics connection: some pedagogical remarks in response to Neuenschwander's question. *Electronic Journal of Differential Equations* Conference 04, 207–213.

Woodward, J. (2003) *Making Things Happen: A Theory of Causal Explanation.* Oxford: Oxford University Press.

Zee, A. (2010) *Quantum Field Theory in a Nutshell.* Princeton, NJ: Princeton University Press.

Zhang, S. (1992) The Chern–Simons–Landau–Ginzburg theory of the fractional quantum hall effect. *International Journal of Modern Physics B* 6, 25–58.

Index

Notes
vs. indicates a comparison
Abbreviations used in index
AQFT - axiomatic and algebraic quantum field theories
CD - cluster decomposition
CGMA - condition of geometric modular action
CMG - condition of geometric action for the modular groups
GQFTs - Galilei-invariant quantum field theory
LC - local commutativity
MC - modular covariance
MPCT - modular P_1 CT symmetry
NQFTs - non relativistic quantum field theories
NQM - non-relativistic, non-field-theoretic quantum mechanics
RCFT - relativistic classical field theories
RLI - restricted Lorentz invariance
SSC - spin–statistics connection
StLC - statistics–locality connection

A

additivity 103
algebraic approach to NQFTs 102–105
  axioms 103t
  restricted Poincaré invariance 103
algebraic approach to RQFTs 6, 8, 24–27, 144, 157
  CPT observables 15
  CPT theorem 28t
  DHR representations 25–27
  Haag–Araki axioms 103t
  MC 25
  sense of locality 49–50
  SSC formulation 28t
  Wightman axiomatic approach to RQFTs *vs.* 27, 28t, 29t
algebraic causality 25, 47, 50, 54, 103
algebraic CPT theorem 134
antimatter 140
AQFTs *see* axiomatic and algebraic quantum field theories (AQFTs)
Araki, H 117
Arntzenius, F 72
asymptotically free theory 37
asymptotically safe theory 37
axiomatic and algebraic quantum field theories (AQFTs) 29, 35–36
  Lagrangian approach to RQFTs *vs.* 54
axiomatic cluster decomposition 60–61
axiomatic NQFTs 88–97
  Lévy-Leblond- *vs.* Wightman- axioms 88–93, 89t
  vacuum state 89–90
  Wightman's approach *see* Wightman axiomatic quantum field theory
axiomatic proof of spin–statistics theorem 173–176
axiomatic theorem, renaming by Greenberg 54

B

Bacry, H 110
Baker, D 39
Bardeen–Cooper–Schrieffer (BCS) theory 142
BEC *see* Bose–Einstein condensates (BEC)
Bell, J
  classical PT theorem 130–131, 132–133
  Lagrangian approach in CPT invariance in RCFTs 130
Benevides, C 134–135
Berger, M 71
Berry, M 121
Bisognano, J
  MC 62
  MPCT 62
Blaylock, G 166
Bohmian dynamics 166
Bokulich, A 153–154
Borel summable divergent asymptotic series 37
Bose–Einstein condensates (BEC) 142
  explanations of 4
Bose–Einstein statistics 10, 41, 56, 141–142
  causal account 151
  integer-spin fields 23
  SSC definition 1
  SSC derivations in NQFTs 121
Bose–Fermi alternative 8
  SSC derivations in NQFTs 121
bosonic effective forces 166
bosons 3
Bronstein cube 105–106, 106f
  problems with 106–108
Bronstein hypercube 108–109, 108f
Brown, H
  Bohmian dynamics 166
  NQFTs and non-relativistic limit 113
Buchholz, O 80
Bueno, O 153, 156

C

Casmir invariants 87–88
Caulton, A 3
causal account 148–152
causal history approach of Lewis 165
causal separation, Minkowski spacetime 96
causality 7–8, 56–57
  CD relation 47
  Lagrangian approach to RQFTs 23, 47, 56
  LC 56–57
  LC relations 56–57
  locality 50
  relations 56–57
  SpLC relations 56–57
  StLC relations 56–57
  violation of 42
causality constraint (microcausality) 22
CD *see* cluster decomposition (CD)
CGMA *see* condition of geometric modular action (CGMA)
Chern–Simons gauge field 142
Chern–Simons interaction 142–143
classical PT theorem, Bell 132–133
classical spacetime 85–86
Clifton, R 153
cluster decomposition (CD) 47, 57–61
  axiomatic version 60–61
  causality relation 47
  justifications for 48
  kinematic *vs.* dynamical constraint 61
  LC and causality *vs.* 61
  locality 50
  and RLI 59–61
  Weinberg's approach 20–21, 47, 97
CMG *see* condition of geometric action for the modular groups (CMG)
compatibility conditions, classical spacetime 86
condition of geometric action for the modular groups (CMG) 62
  fundamentality of constraints on modular data 67, 70–71
  Poincaré covariance from 64–66
condition of geometric modular action (CGMA) 62, 81
  definition by Bucholz 80
  fundamentality of constraints on modular data 67, 70–71
  Poincaré covariance from 64–66
configuration space
  SSC derivation in NQFTs 121–126
  SSC derivation in NQM 8, 10–11, 122–126
conjuctions, deductive–nomological account 146
conjugation symmetry, Lagrangian approach 131
conserved charge 42
convergence problem 33
  purist (rigorous) *vs.* pragmatic (heuristic) approaches 35–36
couplings, purist (rigorous) *vs.* pragmatic (heuristic) approaches 35
CPT invariance 71–77
  alternative theorem formulations 28t
  constraints on 149–150
  definition 1
  derivation 8–9
  extension of 76
  Greenberg's argument 72–77
  observables 15
  RCFTs *see* relativistic classical field theories (RCFTs)
  representation in RQFTs 13–15
  Roberts 10
crossing symmetry 128
cutoff quantum field theory (CQFT) 29, 36
cyclicity of the vacuum 117

D

deductive–nomological (DN) account 145–147
  problem of conjuctions 146
DeWitt, C 121
DHR representations 25–27
Dirac equation 141
DN *see* deductive–nomological (DN) account
Dolpicher, S 43
Dorato, M 152
Duck, I 126–127
  CD justification 48
  Haag's theorem 45
  LC and causality 48
  scattering interactions in CD 58–59
  SSC explanation 143
dynamical constraint, CD 61

E

Earman, J 49–50
empirical import, problem of *see* problem of empirical import
entailment, Greenberg 52–56
equality of masses 140–141
exclusion principle 10
existence problem 36–38
  NQFTs and kinematical intertheoretic relation 113
  pragmatic (heuristic) approaches 74

F

Felline, L 152
Fermi-Dirac statistics 10, 41, 121, 141, 142–143
  causal account 151
  half-integer spin fields 23
  SSC definition 1
fermionic effective forces 166
fermions 3
Feynman, R 143
Fock space 15–16
Fock space operator expression 79–80
fractional quantum Hall liquid 164
  Hall liquid 142

Fraser, D
  AQFT 35–36
  non-relativistic LC 119
French, S 153
fundamentality of constraints on modular data 67–71
  CGMA 70–71
  CMG 70–71
  MC 67–69
  MPCT 69–70

G

Galilei group 95
  irreducible representations 87–88
Galilei-invariant quantum field theories (GQFTs) 7, 87
  axiomatic proof failure 92–97
  failure of separating corollary 96–97
  Lévy-Leblond views *see* Lévy-Leblond, J-M
  PT invariance for vacuum expectation values 93–95
  Wightman axiomatic formulation of RQFTs *vs.* 83
Gell-Mann–Low formula 33, 74, 76
general PT/CT theorem 132
general relativity (GR) 157–158
  Bronstein cube 103–104
  characterization of 167–168
  inertial- and gravitational mass equivalence 158
geometric modular action 61–71
  see also fundamentality of constraints on modular data; Poincaré covariance
GQFT *see* Galilei-invariant quantum field theories (GQFTs)
Greaves, H
  CPT invariance 72
  geometric explanation of CPT invariance 154–155
  Lagrangian approach in CPT invariance in RCFTs 130, 131
  Thomas's general PT/CT theorem 132
Greenberg, O
  algebraic approach renaming 54
  causality and LC 55–56
  CPT invariance 71, 72–77
  entailment 52–56
  interacting RQFT 75–76
  Lagrangian approach renaming 54
  LC 51–52
  Luders–Zumino and Burgoyne proof 79
  non-interacting RQFT 75–76
  spin–locality theorem 55
  spin–statistics theorem 55, 79
  SpLC synonymous with SSC 55–56
  Weinberg's approach 79
  Wightman reconstruction theorem 78–79
Guido, D
  algebraic CPT theorem 134
  MPCT derivation 69
  see also algebraic approach to RQFTs

H

Haag–Araki axioms 103t
Haag–Hall–Wightman (HHW) theorem 114–115
Haag, R
  algebraic version of causality 47
  CD justification 49
Haag's theorem 45
  existence problem 113–114
  Streit–Emch version 119
Hagar, A 72
Hall liquid 142
Halvorson, A 118
helium-3 142
helium-4 142
Hempel, C 146
Hermitian field basis 126–127
Hermitian Lagrangian density 24
heuristic approaches *see* pragmatic (heuristic) approaches
HHW (Haag–Hall–Wightman) theorem 114–115
Hilbert space 14–15
Holland, P
  Bohmian dynamics 166
  NQFTs and kinematical intertheoretic relation 113
Hughes, RIG 153

I

independence conditions, sense of locality 49, 50
indistinguishability 2
interacting Galilei-invariant quantum field theories, non-interacting RQFTs relation 113–114, 160, 161f
interacting relativistic quantum field theories, CPT invariance/SSC non-derivation 157
interacting relativistic quantum field theories, pragmatic (heuristic) and purist (rigorous) approaches 73t
interaction Hamilton density 41
interaction picture 44
intrinsic configuration space 11

J

Jordan product 167
Jost point 41
Jost's axiomatic CPT theorem 74–75

K

Kaku, M 42
  LC and causality 48
kinematic constraints 150
  CD 61

kinematic constraints (*continued*)
  non-interacting RQFTs 161–162, 161f
kinematical aspects of a theory 78
Kitcher, P 147–148
Klein–Gordon equation 92, 141
Kubo–Martin–Schwinger (KMS) state 67–68
Kuckert, B 125–126
  importance of SSC 141
  MPCT derivation 69

L

Lagrangian approach to RQFTs 6, 7–8, 22–24
  AQFT *vs.* 54
  causality 47, 56
  causality constraint (microcausality) 22
  CPT invariance 14
  CPT observables 15
  CPT theorem 28t, 132, 133–134
  renaming by Greenberg 54
  RLI 131
  sense of locality 50
  SSC formulation 28t
  Weinberg's approach *vs.* 27, 28t, 29t
Lagrangian density 126
Lagrangian quantum field theory 144, 157
  conjugation symmetry 131
  CPT invariance in RCFTs 130–133
  NQFTs 83–84, 100–102, 126–129
  RQFTs *see* Lagrangian approach to RQFTs
  speed–space contraction 112
Laidlaw, M 121
Landsman, NP 124–125
Lange, M 153, 156
LC *see* local commutativity (LC)
Lehmann, H 31–32
Leinaas, J 121
leptons 3
Lévy-Leblond axioms 93, 117
  Wightman axioms *vs.* 88–93
Lévy-Leblond, J-M
  GQFTs 7, 88, 90–91
    Wightman axiomatic formulation of RQFTs *vs.* 83
  Haag's theorem avoidance by GQFTs 115
  Poincare group 115–116
  speed–space contraction definition 110
Lewis, D
  causal history approach 165
  PEP 166
Liberati, S 71
Lie product 167
local commutativity (LC) 47, 50–56
  causality 56–57
  locality 50
  RLI relation 47, 51
  StLC *vs.* 16–17, 51
locality
  constraints 50
  as independent condition 49
  localization 49–50
  notion of 6–7
  RQFT 2
  senses of 49
localization 49–50
Longo, R
  algebraic CPT theorem 134
  MPCT derivation 69
  see also algebraic approach to RQFTs
Lorentz group 13
Lorentz invariance (LI)
  restricted *see* restricted Lorentz invariance (RLI)
  violation of 47
Lorentzian spacetime, RQFT *vs.* NQFT 84–85
LSZ reduction formula 31–33
Luders–Zumino and Burgoyne proof 79

M

masses, equality of 140–141
Maxwellian spacetime 86–87
MC *see* modular covariance (MC)
Mickelsson, J 134
microcausality (causality constraint) 22
Minkowski spacetime 13–14, 84–85, 167
  causal separation 96
modular covariance (MC) 47, 62, 103, 145
  algebraic approach to RQFTs 25
  fundamentality of constraints on modular data 67–69
  Poincaré covariance from 63–64
modular data, constraint fundamentality *see* fundamentality of constraints on modular data
modular P_1 CT symmetry (MPCT) 62
  fundamentality of constraints on modular data 67, 69–70
  Poincaré covariance from 63–64
modular stability 66
momentum conservation constraint, scattering interactions in CD 58–59
MPCT *see* modular P_1 CT symmetry (MPCT)
Mullin, W 166
multi-particle states 141
Myrheim, J 121

N

neo-Newtonian spacetime, Maxwellian spacetime *vs.* 86–87
net continuity 66
Newton–Cartan gravity 118–119
  inertial- and gravitational mass equivalence 158
Newtonian gravity 158
NCR *see* normal commutation relations (NCR)

no superluminal propagation (NSP) 49, 50
no superluminal signaling (NSS) 49, 50
Noether's theorem 14
non relativistic quantum field theories (NQFTs) 1, 47
  algebraic approach *see* algebraic approach to NQFTs
  Bronstein cube problems 108
  formulations of 7
  intertheoretic relations 105–116
    see also Bronstein hypercube; Poincaré group
  kinematical intertheoretic relation 112–116
  Lagrangian approach 83–84, 100–102
  RQFT *vs.* 84–88
  SSC 120, 121–129, 158–159, 159f
    configuration space approaches 121–126
    Sudarshan's approaches 126–129
  Weinberg's approach 83, 98–100
non-interacting models 30
non-interacting relativistic quantum field theories
  CPT invariance/SSC derivation 157
  interacting GQFTs relation 113–114, 160, 161f
  kinematic constraints 161–162, 161f
  SSC essentiality 160
non-relativistic causality, Lagrangian NQFTs 102
non-relativistic local commutativity 96–97, 98
non-relativistic, non-field-theoretic quantum mechanics (NQM) 1, 47, 82–119
  SSC 10, 158–159, 159f
non-renormalizable theories 34
normal commutation relations (NCR) 103, 104, 168
  DHR representations 26–27
normal-ordered Lagrangian density 24
NQFT *see* non relativistic quantum field theories (NQFTs)
NQM *see* non-relativistic, non-field-theoretic quantum mechanics (NQM)
NSP (no superluminal propagation) 49, 50
NSpLC (wrong spin–locality connection) 19, 56, 56t, 174
NSS (no superluminal signaling) 49, 50
NSSC (wrong spin–statistics connection) 56, 56t
NStLC (wrong statistics–locality connection) 56, 56t

P

particle identicality 122
particle indistinguishability 122
particle interpretation, Weinberg's approach to RQFT 20
Pauli matrices 117
Pauli, W
  causality constraint (microcausality) 22
  exclusion principle 141–142, 166
  LC and causality 47
PEP (Pauli exclusion principle) 141–142, 166
pertubation theory, Weinberg's approach to RQFT 20
Peshkin, M 125
  LC and causality 48
Pexton, M 165
Poincaré covariance 62–66
  from CGMA and CMG 64–66
  from MC and MPCT 63–64
Poincaré group
  contractions 119
  Galilei group relations 111–112
  irreducible representations 87
  Minkowski spacetime 84–85
  speed'space contraction 109–112
  speed–time contraction 160–161
Poincaré-invariant quantum field theories 87
pragmatic (heuristic) approaches 30–31
  CPT theorem proofs 72
  existence problem 36–38
  Greenberg's argument on CPT invariance 76–77
  interacting RQFTs 73t
  purist (rigorous) approaches *vs.* 2, 34–36
  renormalization problem 31–34
    see also Lagrangian approach to RQFTs; Weinberg, Steven
problem of conjuctions, deductive–nomological (DN) account 146
problem of empirical import 29–30
  purist (rigorous) *vs.* pragmatic (heuristic) approaches 36
Puccini, G 129
purist (rigorous) approaches 29, 144, 157
  CPT theorem proofs 72
  existence problem 36–38, 113
  Greenberg's argument on CPT invariance 73
  interacting RQFTs 73t
  pragmatic (heuristic) approaches *vs.* 2, 34–36

Q

Q-conjugation 131–132
quarks 3

R

Railton, P 166–167
realistic interacting models 30
Reeh–Schlieder theorem 104
relativistic classical field theories (RCFTs) 120, 130–134
  alternative approaches 133–134
  CPT derivation 8–9
  Lagrangian approach 130–133
relativistic quantum field theories (RQFTs) 1, 12–15
  algebraic approach *see* algebraic approach to RQFTs
  CPT invariance representation 13–15
  formulation approaches 6
  interacting 73t, 157
  Lagrangian approach *see* Lagrangian approach to RQFTs
  locality 2
  NQFT *vs.* 84–88
  relativity 2
  representation comparisons 27, 28t, 29, 29t
  SSC *see* spin–statistics connection (SSC)
  Weinberg's approach *see* Weinberg, Steven
relativity 6–7, 47–81
  CPT invariance *see* CPT invariance
  definitions 47, 77
  geometric modular action 61–71
    see also fundamentality of constraints on modular data; Poincaré covariance
  RQFT 2
    see also causality; cluster decomposition (CD); local commutativity (LC); locality
renormalization problem
  pragmatic (heuristic) approaches 31–34
  purist (rigorous) *vs.* pragmatic (heuristic) approaches 35
restricted Lorentz invariance (RLI) 6, 145
  and CD 59–61
  Lagrangian approach in CPT invariance in RCFTs 131
  LC relation 47, 51
  NQFT 82–83
  relations 93–94
  relativity definition 77
  Weinberg's approach to RQFT 20
  Wightman axiomatic approach to RQFTs 18
  Wightman functions as 74
restricted Poincaré invariance 103
Reyes-Lega, A 126, 134–135
rigorous approaches *see* purist (rigorous) approaches
RLI *see* restricted Lorentz invariance (RLI)
Robbins, J 121
Roberts, B
  CPT invariance 10
  T-violation 2

S

S-matrix 43–44
  elements 73
S-principle 128
Saatsi, J 165
SC *see* spectrum condition (SC)
Schrödinger equation 141, 164
  non-relativistic fields 126–127
  second-quantizing 117
Schroeder, D 48
separability, sense of locality 49, 50
separating corollary of spin–statistics theorem 96, 97,173–176
Shaji, A 143
single-particle states 141
Skow, B 152
Sozzi, M 71
spacelike separation 96
spatial constraint, scattering interactions in CD 58–59
spatial local algebras, non-relativistic vacuum 97
spatiotemporal local algebras, non-relativistic vacuum 97
spectrum condition (SC) 81, 145
  Lagrangian approach in CPT invariance in RCFTs 131
  Wightman axiomatic approach to RQFTs 18
speed–space contraction 110–111
speed–time contraction, Poincaré group 160–161
Spekkens, R 151
spin–locality connection (SpLC) 16–17, 56
  causality relations 56–57
  connections 56, 56t
  synonymous with SSC 55–56
  wrong spin–locality connection 19, 56, 174
spin–locality theorem, Greenberg 55
spin–statistics connection (SSC)
  alternative formulations 28t
  connections 56, 56t
  constraints on 149–150
  definition 1
  derivation
    evaluation 8
    in NQM 122–126
  explanatory power 3–4
  failure in Weinberg's approach in GQFT 97
  importance of 141–142
  non-interacting RQFTs 160
  non-relativistic interacting physical systems 142
  NQFTs *see* non relativistic quantum field theories (NQFTs)

NQM 10
physicists' view of 3
RQFTs
explanation by 143–144
representation in 15–17
synonymous with SpLC 55–56
wrong spin–statistics connection 56, 56t
spin–statistics theorem 143
axiomatic proof of 173–176
Greenberg 55, 79
Lagrangian NQFTs 100–101
separating corollary of 173–176
spin, representation in RQFTs 17
SpLC *see* spin–locality connection (SpLC)
standard localization, Halvorson 118
states, representation in RQFTs 14–15
statistics–locality connection (StLC) 16, 43, 56, 77
causality relations 56–57
connections 56, 56t
LC *vs.* 16–17, 51
wrong statistics–locality connection 56, 56t
statistics, representation in RQFTs 15–17
Sterman, G 22
Strevens, M
causal account 149
kairetic account of regulatory explanation 165
StLC *see* statistics–locality connection (StLC)
Streater, R
CD justification 49
representation of CPT invariance in RQFTs 13
separating corollary of spin–statistics theorem 174
Streit–Emch version of Haag's theorem 119
Strocchi, F 49
structural explanation 152–157
geometric explanations of CPT 154–156
problems with 156–157
Sudarshan, E
fundamental theorem 128–129
SSC derivation in NQFTs 8, 126–129
superconductors 4
superfluids 4, 142
Swanson, N 154, 155–156
symmetry groups, representation in RQFTs 13–14
symmetry, crossing 128

T

T transformation 118
T-violation 2
Thomas, T
general PT/CT theorem 132
Lagrangian approach in CPT invariance in RCFTs 130, 131
time reversal invariance 2
Tomita–Takesaki theorem 103–104
algebraic CPT theorem 134
von Neumann algebra 62–63

U

unifying explanation 140, 147–148
unrealistic interacting models 30
UV fixed point, existence problem 37
UV problem 33–34
purist (rigorous) *vs.* pragmatic (heuristic) approaches 35

V

Vacuum Einstein spacetime 85
vacuum state, axiomatic NFQTs 89–90
vacuum, cyclicity of 117
Valente, G 49–50
von Neumann algebra
algebraic approach to RQFTs 25
Tomita–Takesaki theorem 62–63
Vucetich, H 129

W

Wallace, D
cutoff quantum field theory 29, 36
Haag's theorem 45
Lagrangian approach in CPT invariance in RCFTs 131–132
weak local commutativity (WLC) 51, 92–93
Wightman axiomatic approach to RQFTs 19
Weatherall, J
puzzleball view of theories 147
inertial- and gravitational mass equivalence 157–158, 158f
Weinberg, Steven 12, 20–22, 144, 157
CD 20–21, 47, 97
justification 48
CPT 15, 28t
Greenberg, views on 79
Lagrangian approach to RQFTs *vs.* 27, 28t, 29t
LC 42
causality, and 48
NQFTs 7, 83, 98–100
particle interpretation 20
pertubation theory 20
RLI of the S-matrix 20
RQFTs 6, 14
speed–space contraction 112
SSC formulation 28t
Wightman functions 41, 51
Wigner theorem, Hilbert space 14–15
WLC *see* weak local commutativity (WLC)
Woodward, J
causal account 149
explanations of regularities 165

Wightman axiomatic quantum field theory 6, 10, 18–19, 144, 157
- algebraic approach to RQFTs *vs.* 27, 28t, 29t
- CD justification 49
- CPT invariance in RQFTs 13
- CPT observables 15
- CPT theorem 28t
- GQFTs *vs.* 83
- LC 47
- reconstruction theorem 78–79
- RLI 18
- separating corollary of spin–statistics theorem 174
- SSC formulation 28t
- Wightman functions 51

Wightman axioms 40
- Lévy-Leblond axioms *vs.* 88–93

Wightman functions, as RLI 74

wrong spin–locality connection (NSpLC) 19, 56, 174

wrong spin–statistics connection (NSSC) 56, 56t

wrong statistics–locality connection (NStLC) 56, 56t

Z

Zee, A 143

Zimmerman, W 31–32